S0-DUV-456

ONTOGENY AND SYSTEMATICS

ONTOGENY AND SYSTEMATICS

C. J. Humphries,
Editor

Columbia University Press
New York 1988

Library of Congress Cataloging-in-Publication Data
Ontogeny and systematics.
Based on the proceedings of a symposium held in Brighton, Sussex, England, on July 5, 1985.
Includes bibliographies and index.
1. Ontogeny—Congresses. 2. Biology—Classification Congresses.—3. Evolution—Congresses. I. Humphries, C.J.
QH491.0575 1988 575 87-23830
ISBN 0-231-06370-9 (alk. paper)

Columbia University Press
New York Guildford, Surrey

Printed in the United States of America

Clothbound edition of Columbia University Press books are Smyth-sewn and printed on permanent and durable acid-free paper

Book design by Ken Venezio

Contents

Introduction

Five of the essays included here were originally presented as part of the proceedings of a symposium convened at the Third International Congress of Systematic and Evolutionary Biology, Brighton, Sussex, England, on July 5, 1985. Two additional contributions, on the same theme, have been added to this volume.

Ontogeny is the term used to describe the life cycle for the creation of adult organisms from fertilized eggs and is the entire sequence of events involved in the development of an individual organism. Different kingdoms, phyla, and orders of animals and plants are characterized by having different life cycles and phylogeny concerns the changes to which ontogeny has been subject in the course of time. During the last twenty-five years or so there has been a tremendous upheaval in methods for the phylogenetic classification of organisms and for some of the broader principles of classification: these days there is some agreement that such principles as character congruence and parsimony can be used as criteria for choosing between competing classifications. Ontogeny and phylogeny are concepts concerned with the origin and history of living things. Ontogeny is usually considered as quite a separate aspect of the development of single organisms and much less as the consequence of changes in developmental processes which have occurred during evolution. For the most part studies of phylogeny and consequent classifications have relied on comparative anatomy of fossil and modern organisms with little concern for the more dynamic aspects. There is an obvious connection between ontogeny and phylogeny and the aim of this book is to explore the principles of both fields in an effort to combine them.

There is a myth amongst palaeontologists, and indeed among many neontologists, that fossils are of cardinal importance in the reconstruction of phylogenetic history. Naturally we gain important information from studying the fossil record and indeed it is the only information available for extinct groups. However, this gives us only fragmentary and imprecise statements about phylogeny and usually nothing about ontogeny. By concentrating on living organisms we get a much better picture and a more complete idea about evolutionary changes throughout the whole life cycle. The recent revival in systematics of whole life cycles has its origins in the contributions of Nelson (1978) and Løvtrup (1974, 1978). Much of the debate regarding the application of systematic methods to ontogenetic problems has been theorectical rather than empirical (e.g., Alberch 1985; Brooks & Wiley 1985; De Queiroz 1985; Kluge 1985; Nelson 1985).

The crux of both Nelson's and Løvtrup's interpretations of ontogeny has been to recognize that Naef's auxiliary criterion of ontogenetic precedence (Hennig 1966) is a descendant of von Baer's biogenetic law (1828). This law assumes that the transformation of a character during ontogeny "recapitulates" the phylogenetic transformation of this character so that the direction from the general (primitive) to the particular (derived) condition can be observed directly in ontogeny. Nelson (1978) restated the biogenetic law into a falsifiable form: "Given an ontogenetic character transformation, from a character observed to be more general to a character observed to be less general, the more general character is primitive and the less general advanced."

Løvtrup (1984) endorsed the connection between ontogeny and phylogeny by stating that ontogeny is the mechanical cause of phylogeny. In other words ontogeny is a mechanical process while phylogeny is a historical phenomenon. Garstang (1922) summarized the need for a formal systematic approach to the study of whole life cycles when he wrote that "The phyletic succession of adults is a product of successive ontogenies. Ontogeny does not recapitulate phylogeny—it creates it."

Taxonomy is about the classification of whole life cycles, a point made so abundantly clear by the botanist Danser (1950). The unprecedented development of phylogenetic methods in the last twenty years, coupled with the view that new insights into the pattern of life will not come from the application of selectionist explanations to genetic data, led me to think that modern systematists interested in developmental questions may have the means to unify vast amounts of confused, unconnected information on embryology, differentiation, and growth.

All of the contributors to this volume were chosen for their conviction to the idea that phylogenetic analysis is important to understand ontogeny and that attempts to formalize classification of life cycles require some principles derived from empirical studies. The main aim of the symposium which comprised the papers of Løvtrup, Blackmore & Crane, Kluge, Mishler, and Roth was to bring together zoologists and botanists to discuss the connection between ontogeny and phylogeny from a broad perspective. Two more contributions, by André and Weston, are included to make a workable text.

The correct identification of homologies is the very basis of comparative anatomy and ontogeny. By way of an introduction, Dr. V. L. Roth describes some of the pitfalls encountered when viewing ontogeny from such different aspects of development biology as genetics and gross morphology. She draws the distinction between iterative and phylogenetic homologies and gives a review of the definition of homology in relation to ontogeny.

Dr. P. Weston examines the role of parsimony in determining character transformations by direct observations of ontogeny. His purpose is to show that Nelson's reformulation of von Baer's biogenetic law is a special case of a more general method justified by the parsimony criterion.

Prof. A. Kluge develops the theme of classifying whole life cycles. He gives a method for describing continuous variation throughout ontogeny toward providing a model representing growth and differentiation. The main purpose of the model is to analyze allometry so as to make explicit variation and shape change in a way that can be utilized in a phylogenetic research program. He examines the model in terms of heterochronic changes during the allometric phases of growth and demonstrates that ontogenetic trajectories can be analyzed so as to expose coupling and uncoupling of developmental events among species. The model is capable of representing terminal and nonterminal additions or deletions and continuous or discrete changes during ontogeny and overcomes the need for arbitrary divisions into stages in phylogenetic analysis.

Drs. S. Blackmore and P. Crane explore the application of ontogenetic data to phylogenetic and systematic problems in palynology. They describe in general terms the major stages of pollen and spore ontogeny and provide a framework for comparing different developmental pathways. They look particularly at structures of dispersal units, of exines, apertures, and the various caveae and appendages found in pollen and spores. In particular, they assess the value of direct (ontogenetic) versus indirect (comparative anatomy) methods of determining phylogenetic patterns.

Dr. B. Mishler assesses the application of the various theoretical formulations of the relationship between ontogeny and systematics with special reference to Bryophytes. He emphasises that plants differ greatly from animals in terms of their nested modular construction. He describes examples of bryophyte development which would suggest that Nelson's (1978) formulation of the biogenetic law is not a universal and hence not the basis of systematics. He concludes instead that comparative anatomy, character congruence, and parsimony are the principles of systematics.

Dr H. André takes up the theme of iterative and phylogenetic homology by reviewing the stase concept in arthropods. He describes the dynamics of development in the successive stages of the very complicated life cycles of mites in an effort to formalize ontogeny in a way that can be utilised in systematics. By assessing some of the work of Garstang and Grandjean and extending the examples to insects he shows that any given ontogeny tends to repeat the form sequences of related ontogenies but different stases can and do evolve more or less independently from one another and give rise to life cycles differing fundamentally in morphology and ecology. He suggests that there are apparent contradictions to the biogenetic law if iterative homologies showing trends for the less general appear before homologies of a more general distribution during ontogeny are mistaken for phylogenetic homologies. He concludes that it is the lack of ontogenetic theories which has prevented the gathering of developmental information rather than the reverse.

Ontogeny is a description of the changes from an egg through to an adult and epigenetics is the study of the mechanisms through which transformation from the egg to the adult takes place. Before an appreciation can be made of the connection between ontogeny and taxonomy one must understand morphogenetic changes through ontogeny. Prof. S. Løvtrup concludes this book with an epigenetic theory of evolution which goes beyond the synthetic neo-Darwinian theory based on Mendelian genetics and in fact comprises several different disciplines. He indicates, by using von Baer's law, that ontogenetic development constitutes the course of evolution and that a comparative study of epigenetic mechanisms responsible for ontogenesis will elucidate the processes by which differences occur.

REFERENCES

Alberch, P. 1985. Problems with the interpretion of developmental sequences. *Syst. Zool.* 34:46–158.

Baer, K. von. 1828. Ueber die Entwickelungsgeschicte der Thiere. Konigsberg Gebruder Korntrager.

Brooks, D. R. & E. O. Wiley. 1985. Theories and methods in different approaches to phylogenetic systematics. *Cladistics* 1:1–111.

Danser, B. H. 1950. A theory of systematics. *Bibl. Biotheor.* 4:117–180.

Garstang, W. 1922. Theory of recapitulation; a critical restatement of the biogenetic law. *J. Linn. Soc. Zool.* 35:81–101.

Hennig, W. 1966. *Phylogenetic Systematics*. Urbana: University of Illinois Press

Kluge, A. 1985. Ontogeny and phylogenetic systematics. *Cladistics* 1:13–27.

Løvtrup, S. 1974. *Epigenetics—a treatise on theoretical biology*. London; John Wiley.

Løvtrup. S. 1978. On von Baerian and Haeckelian recapitulation. *Syst. Zool.* 27:348–352.

Løvtrup, S. 1984. The eternal battle against empiricism. *Rivista Biol.* 77:183–209.

Nelson, G. 1978. Ontogeny, phylogeny, paleontology and the biogenetic law. *Syst. Zool.* 27:324–345.

Nelson, G. 1985. Outgroups and ontogeny. *Cladistics* 1:29–145.

de Queiroz, K. 1985. The ontogenetic method for determining character polarity and its relevance to phylogenetic systematics. *Syst. Zool.* 34:280–299.

About the Contributors

Dr. H. André is affiliated with the Department of Entomology, the Musée royale de l'Afrique Centrale, Tervuren Belgium.

Dr. S. Blackmore is in the Botany Department, British Museum (Natural History) London.

Dr. Peter R. Crane is affiliated with the Department of Geology the Field Museum of Natural History, Chicago.

Dr. Chris Humphries is the Head of Flora Mesoamericana section, Botany Department, British Museum (Natural History), London.

Dr. Arnold G. Kluge is Professor and Curator of the Museum of Zoology, The University of Michigan, Ann Arbor.

Dr. S. Løvtrup, now retired, was formerly Professor of the Department of Zoophysiology, University of Umeå, Sweden.

Dr. B. Mishler is Associate Professor, the Department of Botany, Duke University, Durham, North Carolina.

Dr. V. Louise Roth is Associate Professor, the Department of Zoology, Duke University, Durham, North Carolina.

Dr. P.H. Weston is affiliated with the National Herbarium of New South Wales, Royal Botanic Gardens, Sydney, Australia.

ONTOGENY AND SYSTEMATICS

1. The Biological Basis of Homology

V. Louise Roth

More than a century after Owen defined a homolog as "the same organ in different animals under every variety of form and function" (1848), discussions of homology continue to appear in the biological literature (e.g., Remane 1952, de Beer 1971, Riedl 1978, Patterson 1982, Roth 1984) and even to fill entire symposia (Sattler 1984, Stevens 1984, Tomlinson 1984a, b). It would appear that the title of de Beer's 1971 essay—"Homology, an unsolved problem"—remains an accurate description.

As many authors (Remane 1952, Bock 1973, Simpson 1975, Riedl 1978) have stressed, it is essential to distinguish between the definition of homology, and the criteria employed to recognize its manifestations. Much interesting controversy has centered on methodological issues. Patterson (1982) equated homology with synapomorphy (a conclusion that emerged from his endorsement of cladistic methods). Fitch (1970) has warned that errors in phylogenetic reconstruction result if "paralogous" ("iteratively homologous" in the sense of Ghiselin 1976 and Roth 1984; see below) molecules are mistaken for "orthologous" (phylogenetically homologous) ones. Similarly, Van Valen (1982) demonstrated that serial correspondence (apparent serial homology) can arise between structures with quite different evolutionary histories. Whereas problems of methodology will (and should) continue to be debated in systematic circles—the correct identification of homologs is, after all, an essential part of any systematic scheme that claims to be "natural"—for the purposes of this paper I shall put aside the issue of how homologs are recognized in order, instead, to examine the biological basis of homology.

I have three main objectives. I begin with a definition of homology. I will contend that it is not difficult to subsume a variety of existing definitions of homology under a single, more general one. My intention here is to demonstrate conceptual links, not to blur distinctions or deny the importance of controversies based on the variation upon the common theme. Second, I will focus on the problems de Beer and others have found in tying the concept of homology to biological processes. These difficulties have led some workers to conclude that perhaps ultimately homology is not a useful concept. I disagree, and I submit that the source of the problems is our ignorance about the relationships between molecular genetics, development, and phylogeny. Where difficulties in applying the concept of homology arise, they point *not* to inadequacies in the concept, but to interesting biological problems that await empirical solution. Last, while resolving these empirical problems will provide a wealth of information of tremendous interest to evolutionary biologists, by no means does that imply that systematists must wait around for the Grand Synthesis: in any given group of organisms phylogenetic information may be lost, obscured, or ambiguous, but there are encouraging signs that much information is still accessible. This will not be news to the professional systematist, for it is the assumption (or hypothesis) under which any systematist works.

A DEFINITION

The fundamental property of homology, according to Van Valen (1982) is "continuity of information." Van Valen used this property as the basis of a definition which, if applied specifically to biological cases, would be the following: "homology is a correspondence between two or more characteristics of organisms that is caused by continuity of information." This is, to my mind, the most succinct, comprehensive, and idealogically neutral definition of homology yet proposed. Its beauty lies in its flexibility: the definition can be used by adherents to any school of thought by simply specifying the relevant kind of information. If "information" is defined idealistically (as it might have been by Owen) it refers to a Bauplan, a blueprint, an archetype, an essence, or an idea in the mind of the Creator. According to modern evolutionary views (in contrast), the information in homologs is primarily genetic, and continuity is provided by genealogy: characters in two species are homologous if the same character is present in the common ancestor. In calling characters homologous, we may infer that they have a common cause (Sober 1984). Alternatively, homologs reveal themselves (Patterson 1982:44, 45, 66) by virtue of their concordance with the distributions of other characters. Patterson (1982) has pointed out important ways in which idealistic, evolutionary, cladistic, and phenetic ideas of homology differ, but one should not forget that these notions also *share* an important property. For example, Owen, Darwin, and Remane have their philosophical differences, but they would not differ fundamentally in the sorts of characters they would choose to call homologous. To the extent that this statement is true, there is common ground for discussion even among biologists who differ philosophically.

The modern usage of the terms "homology" and "analogy" is generally traced back to Owen (1843, 1848; Boyden 1943; Ghiselin 1976; Riedl 1978). Owen also gave serial homology its name (Owen 1848). I have argued elsewhere (Roth 1984) that homology should be defined in such a way that it is applicable to comparisons within as well as between individuals or taxa, and I suggested that a distinction between phylogenetic and iterative (which includes serial) homology might then be made. Continuity of information is relevant to both of these kinds of homology.

1. *Phylogenetic homology*—correspondence between characteristics of different organisms or different taxa is the homology of most interest to the systematist. If homology is defined on the basis of continuity of information, and that information is primarily genetic,[1] then the continuity is provided through genealogy—the ancestor-descendant relationship or, for sister-groups, the sharing of a common ancestor.

2. *Iterative homology* (Ghiselin 1976; Roth 1984) is a correspondence between different structures within a single individual. This includes serial homology, various symmetries, any repetition of structure as with leaves or hairs, arthropod limbs or alpha and beta chains of hemoglobin. Serial, and by extension iterative, homology has been dismissed by some as not a legitimate or true form of homology (Boyden 1943, Mayr 1982, Patterson 1982). It is obviously essential to recognize that phylogenetic and iterative homology are distinct, yet it is also important to acknowledge the conceptual and biological relationship between the two. I maintain (Roth 1984; as would some others: e.g., Ghiselin 1976, Moment 1945, Riedl 1978, Van Valen 1982) that this relationship justifies application

of the term "homology" in both instances. Structures[2] within a single individual may correspond because they are manufactured in the same way. In such cases, the genetic information employed in the development of the corresponding structures may be identical. Other correspondences may arise through gene duplication: genes for alpha and beta chains of hemoglobin differ, but are phylogenetically related; a similar kind of relationship holds for arthropod limbs. In either case, a continuity [if not actual identity] of information is responsible for the correspondences.

Patterson (1982) advocated restricting the definition of "homology" to "synapomorphy." I see no point in narrowing the definition of homology, a word of broader connotation, so drastically. The word "synapomorphy" has been precise and unambiguous since its inception, and especially since its explanation in English (Hennig 1966). To insist that "homology" be used interchangeably with "synapomorphy" beclouds the issue. A *comprehensive* definition of homology invoking continuity of information allows one to acknowledge and to explore the conceptual links between phylogenetic and iterative homology (examples given herein, and in Roth 1984). We may still be as precise as we like by specifying "iterative homology," "serial homology," "phylogenetic homology" (or simply "homology" when phylogenetic homology is obviously implied within the context of systematics), "synapomorphy" or any other restriction of the more general term "homology." To equate homology with synapomorphy is analogous to equating Primates with *Homo sapiens*. Let us maintain a clear distinction between inclusive and restrictive concepts.

In the remainder of this essay I address the questions concerning the biological basis of homology that de Beer raised in 1938 and repeated in frustration in 1971.

HOMOLOGY AND BIOLOGY

Problems

Intuitively (and with some rational basis), biologists look to genetics and developmental biology for the foundations of homology. Yet genetics and development can provide inconsistent pictures. Homology becomes an elusive concept when one attempts to tie it to specific biological processes or relationships or mechanisms. De Beer presented his challenges as a series of examples. I summarize these and other more recent examples in outline form below, labelling them either developmental or genetic, and grouping them according to the type of problem they illustrate.[3]

Homologs do not always develop in similar fashion

1. *The embryological sources or precursors of homologs may differ.*

Few biologists would wish to argue against the homology of newt and lizard forelimbs (specified as vertebrate or tetrapod forelimbs), yet different somites contribute to the musculature in newt and lizard, and the innervation of the muscles also indicates an evolutionary transposition of the limb down the length of the body (de Beer 1971).

Animals are sufficiently conservative in which tissues develop from which portions of the embryo that fate-maps have been useful tools in embryology. Nevertheless, Oppenheimer's (1940) catalogue of instances in which the concept of the embryonic germ layer is violated amply demonstrates the nonspecificity of germ layers. To give one example: urodele primordial germ cells are induced in the ectodermal moiety of the blastula, whereas anuran PGC's develop from specific endodermal blastomeres (Nieuwkoop and Sutasurya 1983).

Regenerating vertebrate lenses are known to arise from the tissue of the iris, or to delaminate from the cornea (Horder 1981).

Many of the unique features of the vertebrates are related embryonically to the neural crest, which may be considered a defining characteristic of the subphylum (Northcutt and Gans 1983). The neural crest is precursor to a diverse assortment of structures and tissue types, and the contributions of neural crest differ in head and trunk regions of a single individual. Yet, few craniofacial skeletal and connective tissue types are produced exclusively by the neural crest. In the head, both the paraxial mesoderm and the neural crest are sources of mesenchyme having musculoskeletal potential (while mesenchyme elsewhere in the embryo is mesodermal), and both of these embryonic tissues may contribute to the same skeletal structure (Noden 1982).

2. *The embryological processes that generate homologs may differ.*

Different patterning mechanisms may underlie the development of homologous structures. In insect development, segmentation arises by a variety of means. At one end of the spectrum, exemplified by the Thysanura and some hemimetabolous species, segments arise by proliferation, or budding, in an anterior-posterior sequence. In other groups, typified by some hymenopterans and dipterans, segments arise simultaneously by a subdivision of the germ band. Sander (1983) concluded that the patterning mechanisms responsible for these two extreme types must differ radically, to a point that led him to suggest that the concept of homology has limited utility.

Inductive relationships may differ in close relatives: one species may require an inductive interaction, while a congener may not. De Beer's (1971) example concerned the frog *Rana fusca,* in which the optic cup is required to induce formation of the lens in the overlying ectoderm. If the optic cup is excised, the lens fails to develop. Yet in *Rana esculenta,* lenses develop whether the optic cup is in place or not (see also Horder 1981, 1983, Hall 1983).

Developmental sequences are occasionally altered in related species, and new developmental stages may be interpolated in the evolution of ontogeny. For example, in the evolution of holometabolous insects, it appears that the pupal stage has been interposed between larval and adult portions of the life cycle (Hinton 1963).

Homology cannot be reduced to a simple correspondence between genes

1. *Pleiotropies.*

Single genes can affect phenotypically disparate characteristics, yet characters are not ordinarily considered homologous by virtue of pleiotropic relationships alone. For example, mutation of a single gene (the *Ta* allele) in lab mice can cause a kink in the tail, small eye apertures, and fused vibrissae (Kalter 1980; see also Grüneberg 1943). At the same time, however, systematists are interested in identifying unit characters that behave independently of each other. Patterson (1982:41) computed the probability that two or more homologies specify the same group by chance alone as a product of combinatorial ratios. As any textbook on probability theory would indicate, the calculation of a joint probability as the simple product of individual probabilities rests upon the assumption that the events are mutually independent. One would not assume *a priori* that i) a fusion between the radius and ulna and reduction in the size of the kidney (from the *ld* allele in lab mice), or ii) changes in the shape of the skull, changes in pelvis morphology, modification of the shape of the cerebellar floccular lobe and absence of lachrymal glands (the *fi* allele), would each be affected by a single allele (Kalter 1980). Yet the congruence of sets of characters such as these would be taken as corroborative evidence in a cladogram.

2. *New genes, previously unassociated with the development of a particular structure, can be "deputized" in evolution; that is, brought in to control a previously unrelated developmental process, so that entirely different suites of genes may be responsible for the appearance of the structure in different contexts. We may call this "genetic piracy."*

The ease with which transdetermination or homeotic mutations can convert an antenna into a leg or a haltere into a wing in *Drosophila* appears to be a very satisfying reflection of the iterative homology of these structures (Garcia-Bellido 1977; Raff and Kaufman 1983, chapters 9 and 10). During evolution, halteres were derived from wings, and antennae are believed to be modified legs. Yet it is much less comforting to note that transitions also occur between eyes and wings or antennae and eyes (Hadorn 1965, Garcia-Bellido *et al.* 1979). There has never been any suggestion of homology between these structures. It appears that the structures that arise from imaginal disks in development—legs, wings, eyes, etc.—share certain genetic switch mechanisms, although they may share no phylogenetic relationship to one another. There has been a reorganization of the genome in such a way as to bring unrelated structures under the hierarchical control of one set of genes.

De Beer (1971) cited an example of eyeless mutants in *Drosophila.* With these mutants, it is possible to select for other genes that modify the expression of the eyeless mutation so that the end result is a fly with restored eyes that still has the original mutation. Consequently, new genes are involved in the formation of eyes that previously had not been.

I suggested a similar scenario of genetic or developmental piracy for dental evolution in elephants (Roth 1982) and for the convergent evolution between fore and hind limb morphology in tetrapods (Roth 1984). The similarities in locking mechanisms of dorsal, ventral, and pelvic fin spines in sticklebacks (Hoogland 1951), or dorsal and anal fin spines in

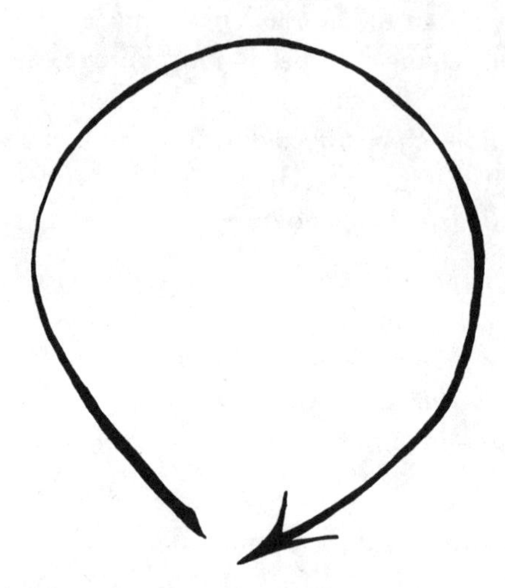

Figure 1.1. A single life cycle.

ponyfishes (Seigel 1982) may be explained by a similar evolutionary process (M. A. Bell, pers. comm.).

From the problems enumerated above, a question arises: what is the relationship between the concept of homology and the processes of genetics and development?

Connections

A full description of the phenotype of an organism comprises an entire life cycle (Bonner 1965, Nelson 1985) (figure 1.1). If we consider more than a single generation, we can illustrate this sequence as a cycle that undergoes translation through time; the result is thus a helix (figure 1.2). In this illustration, corresponding positions on different whorls of the helix are corresponding parts of the life cycle. As Gould (1977), Hennig (1966), and others have emphasized, if *portions* of a life-cycle are to be examined, comparisons are most appropriately made between such corresponding stages in life.

I will choose the zygote as a starting point, although the argument

Figure 1.2. Ontogeny and phylogeny can be represented as a helix: a single life cycle (drawn in fig. 1.1) undergoes translation through evolutionary time. Sexual reproduction would make the illustration more complicated, but one might imagine a population as a large number of helices that form separate strands converging, intersecting at the point of the zygote, and diverging in a meshwork. One can carry a metaphor too far, but the helix may facilitate visualization.

works equally well for spores and other fully competent cells. With the exception of such cases, the zygote is the smallest unit that contains all the information necessary to make an adult organism (Bonner 1974). The unit is a cell and not the genetic material. As Whittle (1983) phrased it: "all cells derive from existing cells and not de novo by assembly from their components"—genetic and cytoplasmic information as well as the cell's machinery are all essential (which is not to say that the environment of the zygote has no hand in structuring the adult phenotype, but merely that genetic and cytoplasmic information within the zygote determines whether it is to develop into a moss or into an elephant). As Bonner (1974) suggested, the zygote may be considered the stage of the life cycle in which there is maximal storage of information, and minimal expression. Ontogeny, then, is the translation, or elaboration, of this stored information into its phenotypic expression. Ontogeny *embodies* the information continued between successive generations, and ontogeny is therefore a *manifestation* of homology.

Then how should we study development? and how do we resolve the problems posed above?

"Ontogeny," as I am using the word, covers a broad spectrum of phenomena. Ontogeny is studied at several levels, by different sorts of biologists: geneticists, developmental physiologists, systematists.[4] A common way of depicting relationships between these different biological realms was illustrated by Alberch (1982) and Oster and Alberch (1982) (figure 1.3), who showed essentially three tiers—genetic, developmental, and phenotypic. Arrows were directed between the tiers in the diagram to indicate possible influences and determinations, or filters (labeled "natural selection" or "developmental parameters") were interposed to indicate processes that constrain the expression of phenomena in successively higher levels. I'd like to suggest that the failure to understand the differences, and the connections, between these different levels is the root of difficulties with tying homology to biological mechanisms. Some understanding will be improved as new empirical information comes to light, but a large part of de Beer's (1938, 1971) and Sander's (1983) dissatisfaction with the concept of homology arises from a failure to recognize the complexity of information already available to us.

Figure 1.3. In diagram (A) (reproduced with permission from Oster and Alberch 1982), an organism is represented by a single point on each plane, while diagram B (reproduced with permission from Alberch 1982) depicts processes taking place within a single organism. In the text I refer to three levels, which may approximately correspond to regions shown in these two diagrams. The lowest level on both diagrams is the realm of genetics; the second level, the realm of developmental biology, is enclosed within a dashed line in (A), whereas in (B) it would be situated somewhere between the "Developmental Parameters" and "Phenotypes" tiers. The third level I would label "gross morphology" or "gross phenotype." The word "phenotype" as it is used in these diagrams is confusing, because proteins and cell properties are aspects of phenotype. I also see difficulties with distinguishing between a "phenotype" and a "realized phenotype," or confining selection to a single level in the diagram.

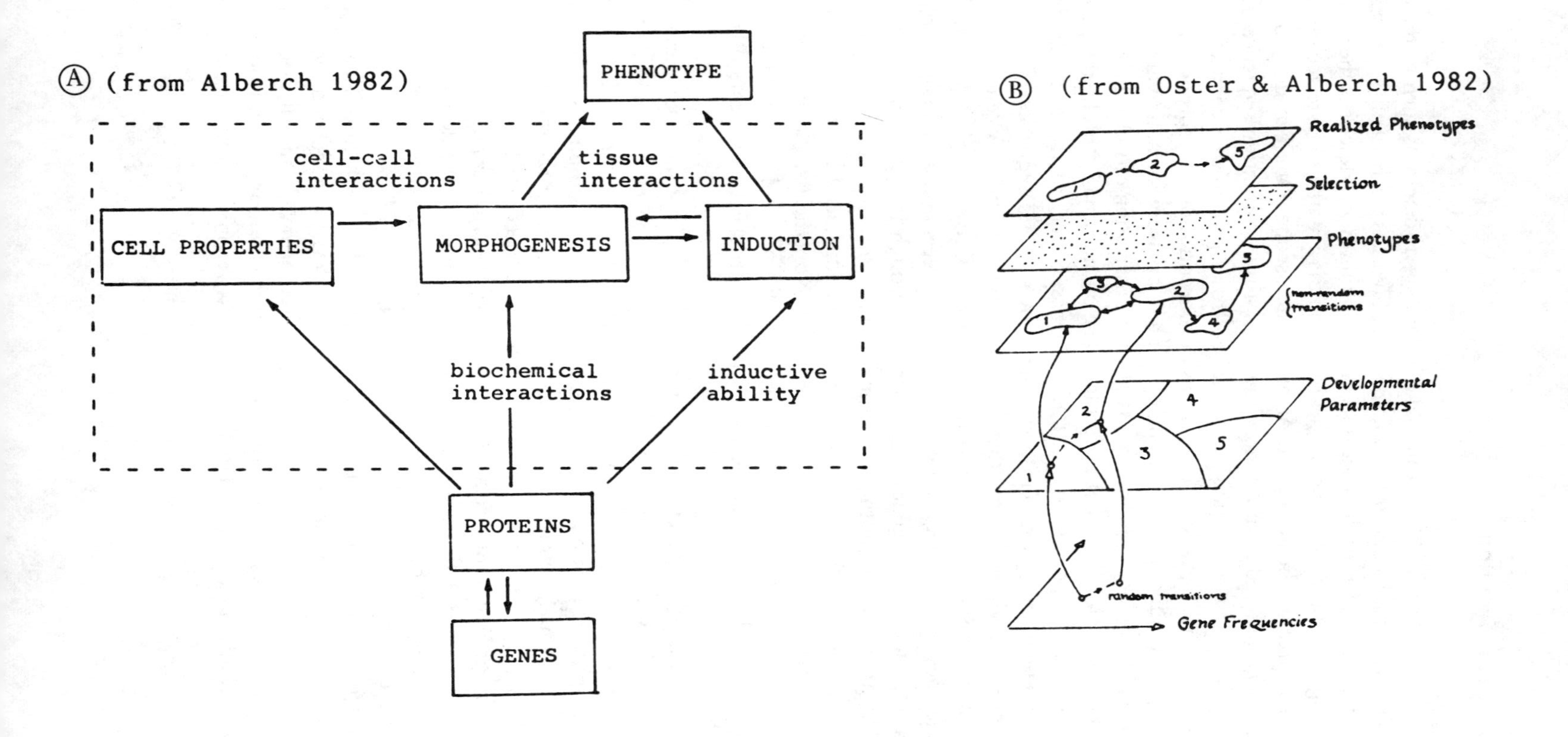
Ⓐ (from Alberch 1982)
PHENOTYPE
cell-cell interactions
tissue interactions
CELL PROPERTIES
MORPHOGENESIS
INDUCTION
biochemical interactions
inductive ability
PROTEINS
GENES
Ⓑ (from Oster & Alberch 1982)
Realized Phenotypes
Selection
Phenotypes
non-random transitions
Developmental Parameters
random transitions
Gene Frequencies

One mistake that one can make in looking at diagrams such as those in figure 1.3 is to interpret the arrows and boxes as a cybernetic *sequence*. In fact, however, each of the tiers drawn in this model describes the same thing. Whichever tier one chooses to examine is a matter of taste, technical repertoire, or the resolution of one's microscope. One can describe a change observed in gross morphology as a transformation sequence (to which certain rules apply, such as Geoffroy's principle of connections, or allometric variation); or in terms of concentrations of morphogens or shapes of individual cells (using principles of pattern formation and concepts like induction); or in terms of the transcription of individual genes (the interaction of sets of molecules). The phenomena in this hierarchy—gross phenotype, development and genetics, or the reverse, genetics, development and gross phenotype—are not a causal sequence. They are embedded within one another, and inseparable as units. A gene product is itself one manifestation of the phenotype. Gross morphology, which includes the relative positions of particular sheets of tissues for example, in turn may affect genetic transcription through inductive interactions. The arrows between the levels run in several directions.

Genetics, development, and gross morphology, as different aspects of the same phenomena, are causally linked. Nonetheless, the way we describe processes at each of these levels is distinct, and what we refer to as a process, or a sequence, or a result, varies with the scale of our observations.[5]

To state that genetics, developmental physiology, and comparative morphology are different fields, with different phenomenology, may sound completely self-evident, but I don't believe that it is. Take, for example, the common tendency to speak of genes as though they coded for individual characters, or worse, for individual structures. My point is that a process as it is described at one level need not bear any simple linear or one-to-one relationship with processes observed at another. Alberch (1985) made a related point in distinguishing developmental sequences from developmental processes[6] and I suspect the problem may be even more pervasive than he suggested.

It is easy to be misled by tantalizing examples in which order *is* preserved in development. For example, one can devise a model of the genetic control of development that is simple, well-behaved, and hierarchical. Figure 1.4 shows a model of ectoderm differentiation, taken from Raff and Kaufman (1983), that depicts successive bifurcations in develop-

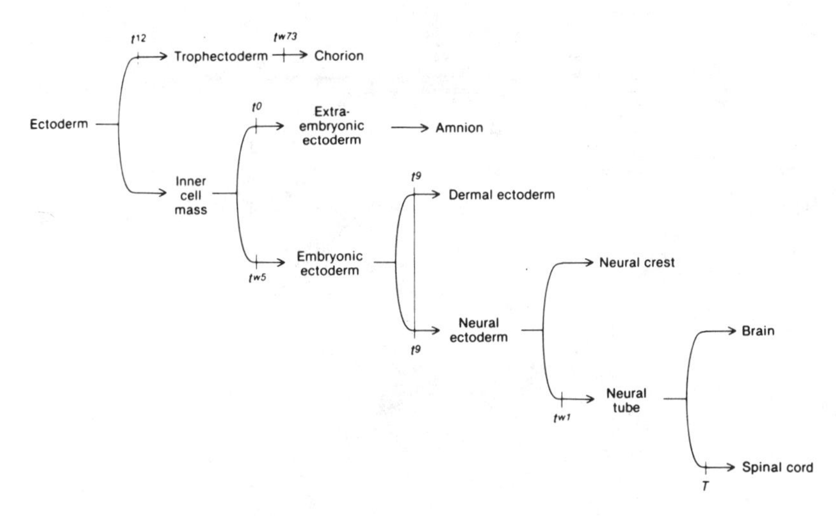

Figure 1.4. A model of ectoderm differentiation. A change in any single allele in this model would produce consequences for the emergent embryological pattern that are predictable and straightforward. Figure reproduced from *Embryos, Genes, and Evolution* by R. A. Raff and T. C. Kaufman 1983. Copyright © 1983 by MacMillan Publishing Company. Reprinted by permission of the publisher.

mental paths. Genes are identified that operate the switch mechanism at each point, in a simple linear sequence. A change in any single allele in this model would produce consequences for the emergent embryological pattern that are predictable and straightforward.

Yet development often follows a more complex path than simple bifurcation. Figure 1.5, taken from Riedl (1978, after Coulombre 1965), illustrates the complex interactions involved in induction and differentiation in the vertebrate eye. Embryology can involve nonlinear relationships, feedback loops, one-to-many and many-to-one relationships. Figure 1.6 shows one of Kauffman's models of the dynamics of gene control, a network of interactions. It becomes topologically impossible to superimpose an embryological sequence upon such a genetic network. This network may have a high degree of order to it and it may produce a highly organized phenotypic product. But in contrast to a simple linear bifurcation model (figure 1.4), the morphological consequences of making a single change in this genetic network are by no means obvious.

In general, there is no necessary *simple* congruence between genetic,

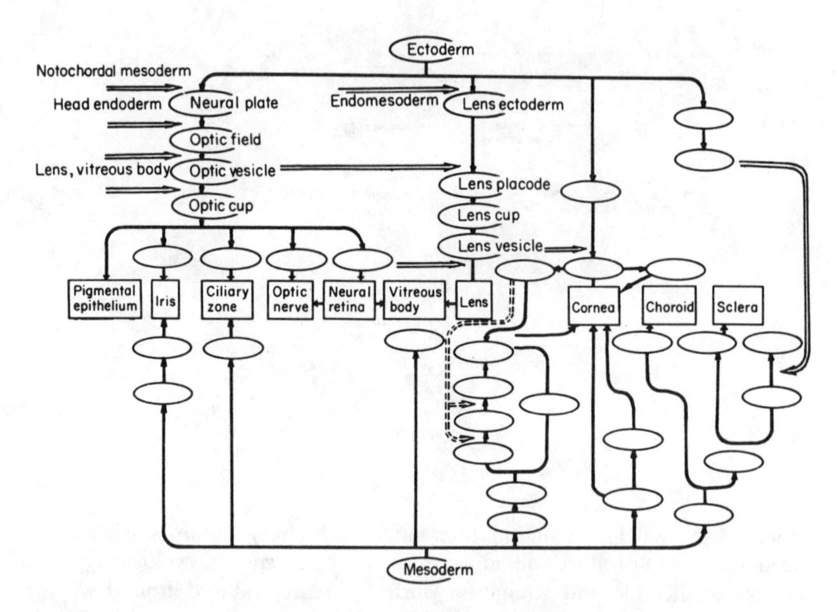

Figure 1.5. The pattern of induction and of differentiation in the vertebrate eye (figure simplified after Coulombre 1965 by Riedl 1978). Arrows represent paths of differentiation, and ellipses (some of which, for simplicity, were left unlabelled) are intermediate embryonic stages. This developmental network shows a higher order of complexity (nonlinear relationships, feedback loops, one–to–many and many–to–one relationships) than that illustrated in fig. 1.4. From Riedl, *Ordnung des Lebendigen*. Copyright © 1975 by Paul Paray, Berlin and Hamburg. Reprinted by permission of the publisher.

developmental, and evolutionary pathways. This fact is at the root of the problems identified by de Beer. It is the reason why it is necessary to distinguish between processes and sequences at any given level, and it is an obstacle to any strict application of the biogenetic law (Nelson 1978, Alberch 1985).

Occasionally, information adduced at one level *will* be useful at others, and this occurs sufficiently often to arouse the unfounded expectation that changes in development are invariably transparent in their simplicity. The gross phenotype sometimes does provide obvious clues to the shifts that have taken place in developmental relationships. Riedl (1977, 1978, 1982) has suggested that there should be natural selection for genetic organization and epigenetic systems that are arranged in a hierarchical

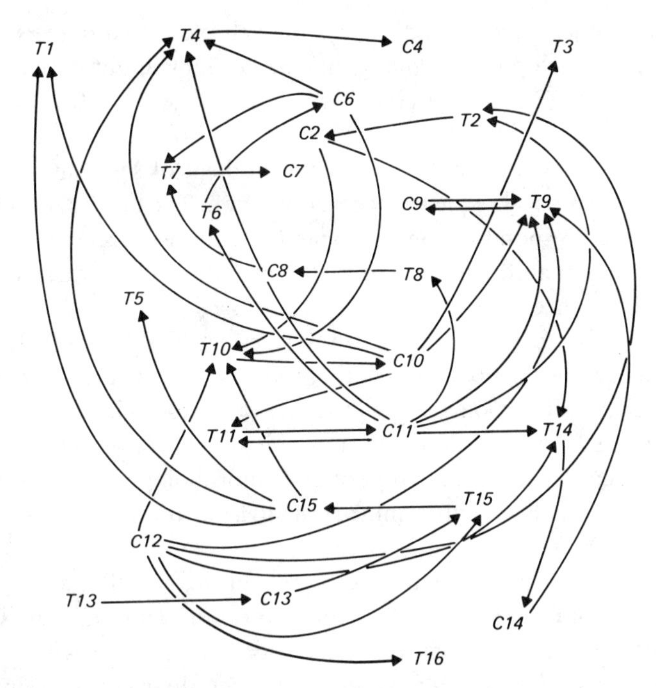

Figure 1.6. An example of a model suggested by Kauffman (1983) for the dynamics of gene control. A genetic network of this order of complexity would not be isomorphic with the network of developmental interactions (observed at a morphological level) that it controls. It might be possible to *deduce* what effect a single mutation in this network would have on the sequence of embryological changes in gross morphology, but the correspondence between a genetic network like this and the embryological stages that emerge from it is not simple. From "Developmental Constraints: Internal Factors in Evolution," by Stuart A. Kauffman, in *Development and Evolution*, edited by B. C. Goodwin, N. Holder and C. C. Wylie. Copyright © 1983 and reproduced by permission of Cambridge University Press.

manner similar to the "phenes" for which they code—that order and correspondences in phenotype should directly mirror genetic architecture.

In cases of simple heterochrony (Gould 1977; Alberch et al. 1979), developmental sequences reflect evolutionary sequences, with no apparent disruption of underlying developmental processes. Thus, the adult of a descendent species of cockle may look like the juvenile form of its ancestor, and data for the Irish elk fall on an extension of the allometric curve for other deer (Gould 1977). In such cases, parameters of size, shape, and developmental timing may be decoupled, but developmental sequence and order are preserved. Segmentation genes in *Drosophila* are another case in point. Their serial arrangement on the chromosome superficially suggests the antero-posterior gradient of segment morphologies, and a temporal sequence of determination. The borders of segments in fact do correspond to compartmental boundaries, and compartments appear to be manifestations of developmental fields, or realms of gene action (Lawrence 1981).

And yet the correspondences are not perfect (Gubb 1985). Relationships and arrangements are not entirely what we would expect from what we typically call homologies.[7] As the title of a Research News feature in *Science* asked; "Why is development so illogical?" (Lewin 1984), following the discovery that symmetries of gross morphology in *Caenorhabditis elegans* are not based on symmetrical division of cellular populations (Sulston et al. 1980).

The answer is that evolution is opportunistic, and that all aspects of the process of ontogeny are vulnerable to evolutionary change. Certain types of alterations of ontogeny may occur or survive with differing frequency (e.g., Arthur 1982; Gould 1977), but there is no reason to *rule out* the possibility of changes in inductive relationships or embryological precursors, or to be surprised when it appears that a particular suite of genes has "taken over" a developmental process over which it previously (in phylogeny) had no influence. Despite the resurgence of interest in evolutionary "dissociations" in the form of heterochrony (Gould 1977; Alberch et al. 1979), a very large class of dissociations has been overlooked. Attention has focused on heterochronic dissociations, while dissociations of the kinds enumerated in table 1.1 remain troublesome, but comparatively unexamined (though see Raff and Kaufman 1983 for some important steps in this direction). The relationships between processes at genetic, developmental, gross phenotypic, and evolutionary levels remain a black box. Examining that black box has traditionally been the job of experimental, rather than comparative biologists, but the contents should be of interest to us all. (see note 4, and Alberch 1985.)

What has this got to do with homology?

Phylogeny has components of both continuity and change. We identify the elements of continuity as homologies; the change is called evolution.

Table 1.1. Evolutionary Dissociations:

—change of embryological source or precursor

—change in embryological processes
- change in patterning mechanisms
- change in inductive relationships
- change in developmental sequences
- heterochrony

—genetic "piracy"

Note: The list is not exhaustive. See text for discussion.

Because processes described at different levels (though causally linked) are not congruent in any simple way, we should not be discouraged when conservatism at one level of description appears to be mirrored in change at another. If homology is to be a useful concept applied to all of these levels, we must be willing to be specific, and explicit, about precisely which features of the life cycle (as it is described at any level) we are describing. It may be adequate for some purposes to speak of a homology between newt and lizard forelimbs, specified as vertebrate limbs. It may ultimately be more precise, and more interesting, however, to point to a conservatism in the pattern-formation mechanism (Goodwin and Trainor 1983), with a change in the number, or relative positions, of somites. We will have more suitable objects for comparison, and avoid some of the ambiguity and confusion biologists have encountered with the concept of homology in the past, if we cease to insist on homologizing entire structures (Roth 1984), and instead decompose structures (or collections of structures; Hubbs 1944) into their individual characteristics (properties, individual features, or aspects of development).[8]

THE ACCESSIBILITY OF PHYLOGENETIC INFORMATION

Since this volume concerns the relationship between ontogeny and systematics, I would like to end with an example, and a note of hope, taken from the work of my colleague Fred Nijhout and his students Greg Wray, Claire Kremen, and Carolyn Teragawa (Nijhout et al 1987). They have constructed a computer model that simulates general aspects of ontogeny. Basically it is a two-dimensional cellular automaton (Lindenmayer 1982,

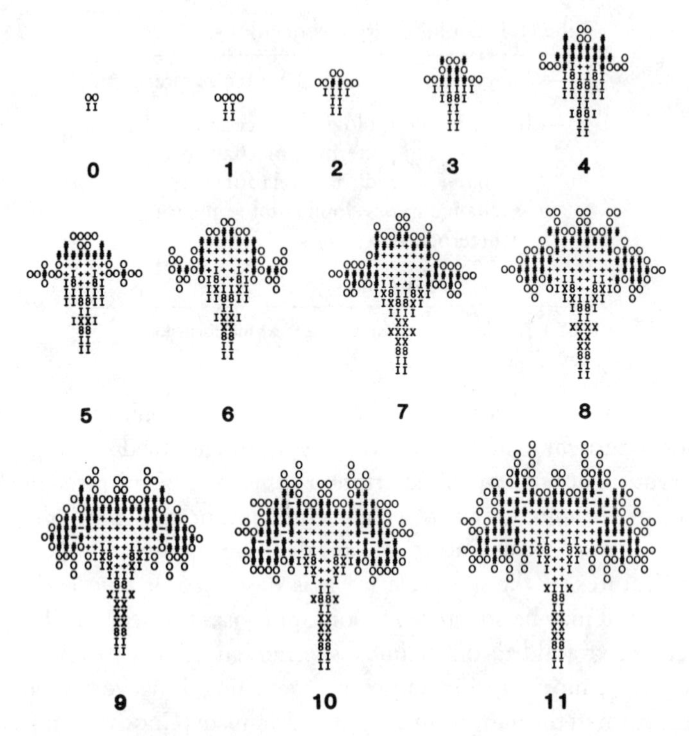

Figure 1.7. Ontogenetic series for the "wild type" pattern of the model described in the text. Numbers refer to ontogenetic stages. Reproduced with the permission of H. F. Nijhout, G. Wray, C. Kremen, and C. Teragawa.

Wolfram 1984) that begins with a polarized, bilaterally symmetrical "zygote," which increases the numbers of its cells, differentiates, undergoes cell death, and produces a morphogen, according to a specified algorithm. Figure 1.7 shows the ontogenetic series for the "wild type" pattern. I should stress that these patterns are simply patterns, and are not intended to represent a specific type of structure or organism.

The model, though of necessity simple by biological standards, has proved useful for examining the relationship between changes in ontogeny and the evolution of morphology. The utility of a model over natural systems in this case arises from the fact that parameters are controlled directly, and assigned in advance rather than inferred. The developmental rules used in the algorithms were limited to known activities of gene

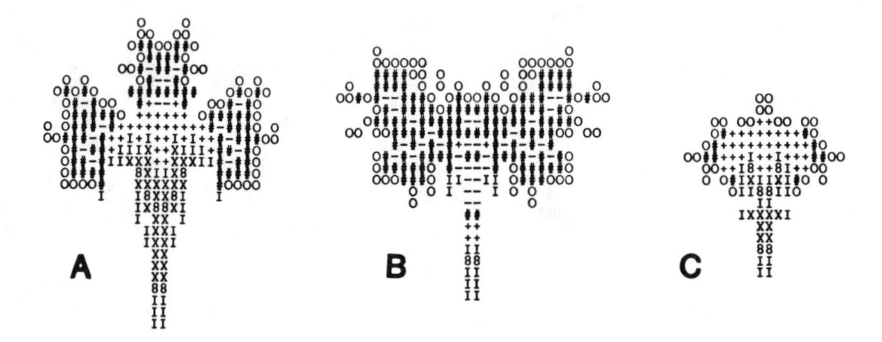

Figure 1.8. Results of introducing single heterochronic mutations in the ontogeny of the pattern shown in fig. 1.7. Pattern A arises if division of cells of type "I" is prolonged through age six, instead of ending at age five as in the original algorithm. If the endpoint of these cell divisions occurs at age four, pattern B results. Pattern C is generated if cells of another type ("O") cease division at age five, rather than seven. Reproduced with the permission of H. F. Nijhout, G. Wray, C. Kremen, and C. Teragawa.

products, and properties known to characterize cells. With the introduction of a series of mutations, or changes in the developmental rules, the model successfully mimics many of the emergent properties of development and of evolution. Indeed, for example, it is observed that genetic parallelisms do not always result in morphological parallelisms. If a mutation is introduced which allows some cell types to continue dividing for a longer period, an example of a heterochrony at the level of cell division, the form illustrated in figure 1.8a results. Figures 1.8b and 1.8c result from other heterochronic mutations, this time involving early cessation of division. Such heterochronies at the level of cellular division produce forms that would not be considered heterochronic on purely gross morphological grounds. The patterns in figure 1.8b and c do not interpolate nicely into the ancestral ontogenetic sequence shown in figure 1.7.

Of particular relevance to this volume will be the results of phylogenetic reconstructions. Sets of five or six terminal phenotypes, such as the patterns at the top of figure 1.9, were shown to an entomologist, two systematic botanists, and an ichthyologist who were ignorant of the underlying phylogeny, and who did not know the rules by which the patterns were generated. They were allowed to identify characters as they

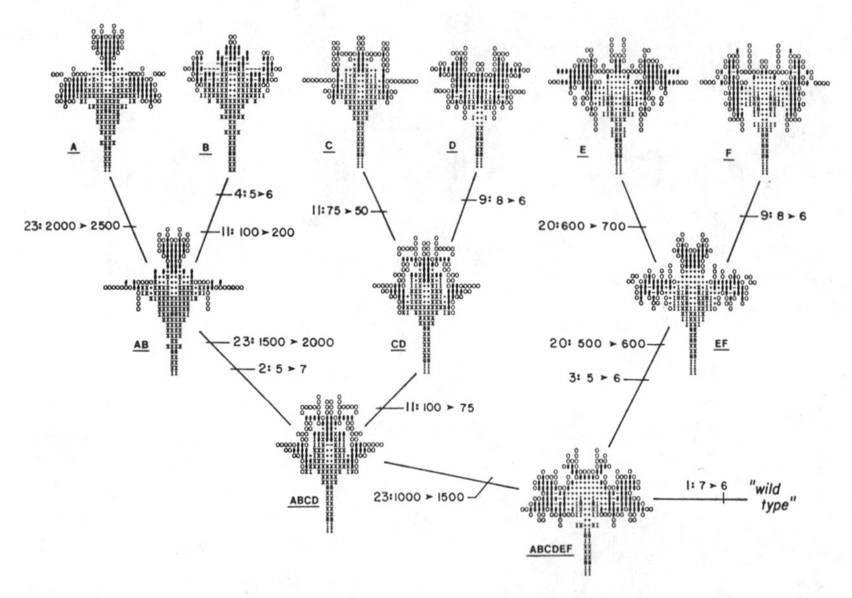

Figure 1.9. A phylogeny of patterns produced by Nijhout et al. (1987 reproduced with permission). Only end-products, or "adult" patterns are shown on this diagram. Numeric codes refer to "mutations," or changes in the pattern-generating algorithm (see fig. 1.8 for a description of three such mutations). A phylogenetic reconstruction of this genealogy placed D as a sister group to (A,B,C) rather than to C alone, but was otherwise correct.

chose, and analyze these characters cladistically, both by hand, and by using the Wagner 78 computer parsimony method. Early stages of ontogeny were available, and characters from these early stages turned out, at times, to be essential for resolving the cladograms.

Even though the mechanisms for generating the patterns were controlled at the level of individual cells, and were not known by the systematists, in all but one case (six independent phylogenies; see figure 1.9 for the exception), the phylogenetic relations deduced on the basis of gross morphology were in agreement, and were correct.

SUMMARY

To conclude, I can make three generalizations.

1. Homology is "correspondence caused by a continuity of informa-

tion" (Van Valen 1982). This definition works for comparisons between taxa, as well as between structures within single individuals. Particular applications of the term may specify the relevant type of information.

2. There is no neat congruence between that information as it is described in genetic, developmental, and gross morphological or evolutionary contexts, even though the fields of genetics, development, and comparative morphology represent only different points of view on the *same* phenomena.

That, however, is not a major problem for the utility of homology as a biological concept. Much attention has focused on heterochrony as a type of evolutionary dissociation or decoupling. We should not be surprised upon encountering other types of evolutionary dissociation, such as genetic piracy, shifts in inductive relationships, or changes in precursors.

3. Despite the dissociations that characterize evolution, and despite a lack of congruence, in a purely topological sense, between different levels of description, the information for reconstructing phylogeny remains accessible for systematic study.

ACKNOWLEDGEMENTS

I thank Fred Nijhout for permission to cite unpublished work, and Claire Kremen, Fred Nijhout, Lucinda McDade, John Lundberg, Tom Colton, Ellen Simms, and Steve Wainwright for challenging discussion and comments on the manuscript. John Gregg, Greg Wray, Leigh Van Valen, and Paula Mabee made constructive suggestions for improvements on the final draft.

NOTES

1. The information needn't be entirely *genic*. As ciliate protozoans nicely illustrate (Sonneborn 1970, Frankel 1983), morphological information can be conveyed on templates other than DNA or RNA.
2. One need not be confined to morphological structures in making comparisons. Although for simplicity I will refer to the objects of comparison as structures, it may be understood that biochemical pathways and even behaviors (Hall and Williams 1983, Eberhard 1982) can have homologs.
3. These examples point to a weakness in my earlier attempt to integrate developmental information with the concept of homology (Roth 1984). In the

previous paper I stated that homologs *must* share developmental pathways, and acknowledged, but did not confront, some of the difficulties with this approach. The current paper addresses these problems. With the exception of this particular point I stand by the other conclusions voiced in Roth (1984).

4. Most of the discussion of the relationship between systematics and ontogeny has taken place among systematists who compare material at the level of gross morphology. We currently have "biochemical systematists" and "morphological systematists" (the latter including those who compare embryological sequences). Perhaps the questions that remain about the relationship between homology and biological processes can be answered by a new variety of systematist who studies comparative developmental physiology.
5. And natural selection operates at all three of these levels, as Sander (1983) illustrated.
6. It might be argued that this distinction is a relative one, and that the relationship is transitive through several levels of description. What appears to be a *process* at one level is actually a *sequence* from a lower tier in a reductionist hierarchy.
7. As complex as the relationships between genes, developmental processes, and morphology appear to be (in what Brenner called "the grammar of development," Lewin 1984), one has hope that these relationships eventually will yield to analysis. Thus Kauffman (1973) produced a persuasive model that interrelates a hypothetical set of genetic addresses (a combinatorial code, or "states of a circuit") with (1) a topographical arrangement of compartmental boundaries in the embryo and (2) patterns and frequencies of transdeterminative events. The model is powerful in accounting for a range of observed phenomena although details of the combinatorial code are inaccurate (Kauffman 1973; Raff and Kaufman 1983:276).
8. The tendency to think in terms of entire structures is strong. Although Wiley (1981), for example, defined "character" in terms of "attributes" of an organism, the examples of homologous characters that followed were entire structures. Decomposing structures into individual attributes that may, or may not be homologized eliminates the need to speak of homology as a matter of degree, which I had suggested earlier (Roth 1984, Van Valen, pers. comm.). If we think of homology as the element of correspondence or continuity (and evolution as the element of change), then we can tabulate the characteristics of two organisms in such a way that each property either does or does not correspond to a property in the other. This exercise will be a familiar one to the systematist.

REFERENCES

Alberch, P. 1982. Developmental constraints in evolutionary processes. In J. T. Bonner, ed. *Evolution and Development,* pp. 313–332. Berlin; Springer-Verlag.

Alberch, P. 1985. Problems with the interpretation of developmental sequences. *Syst. Zool.* 34:46–58.
Alberch, P., S.J. Gould, G. Oster, and D. Wake. 1979. Size and shape in ontogeny and phylogeny. *Paleobiology* 5:296–317.
Arthur, W. 1982. A developmental approach to the problem of variation in evolutionary rates. *Biol. J. Linn. Soc.* 18: 243–261.
Bock, W.J. 1973. Philosophical foundations of classical evolutionary classification. *Syst. Zool.* 22:375–392.
Bonner, J. T. 1965. *Size and Cycle.* Princeton: Princeton University Press.
Bonner, J. T. 1974. *On Development.* Cambridge: Harvard University Press.
Boyden, A. 1943. Homology and analogy: a century after the definitions of "homologue" and "analogue" of Richard Owen. *Quart. Rev. Biol.* 18:228–241.
Coulombre, A. 1965. The eye. In R. DeHaan and H. Ursprung, eds. *Organogenesis,* pp 219–253. New York: Holt.
De Beer, G. R. 1938. Embryology and evolution. In G. R. de Beer, ed. *Evolution; Essays on Aspects of Evolutionary Biology Presented to Professor E. S. Goodrich on his Seventieth Birthday,* pp. 57–78. Oxford: The Clarendon Press.
De Beer, G. R. 1971. *Homology, an Unsolved Problem.* London: Oxford University Press.
Eberhard, W. G. 1982. Behavioral characters for the higher classification of orb-weaving spiders. *Evolution* 36:1067–1095.
Fitch, W. M. 1970. Distinguishing homologous from analogous proteins. *Syst. Zool.* 19:99–113.
Frankel, J. 1983. What are the developmental underpinnings of evolutionary changes in protozoan morphology? In B.C. Goodwin, N. Holder and C.C. Wylie, eds. *Development and Evolution,* pp. 279–314. Cambridge: Cambridge University Press.
Garcia-Bellido, A. 1977. Homeotic and atavic mutations in insects. *Amer Zool.* 17:613–629.
Garcia-Bellido, A., P. A. Lawrence, and G. Morata. 1979. Compartments in animal development. *Sci. Amer.* 241:102–110.
Ghiselin, M.T. 1976. The nomenclature of correspondence: a new look at "Homology" and "Analogy." In R.B. Masterton, W. Hodos, and H. Jerison, eds. *Evolution, Brain and Behavior: Persistent Problems,* pp. 129–142. Hillsdale, N. J.: Lawrence Erlbaum.
Goodwin, B. C. and L. E. H. Trainor. 1983. The ontogeny and phylogeny of the pentadactyl limb. In B.C. Goodwin, N. Holder and C.C. Wylie, eds. *Development and Evolution* pp. 75–98. Cambridge: Cambridge U. Press.
Gould, S. J. 1977. *Ontogeny and Phylogeny.* Cambridge: The Belknap Press of Harvard University Press.
Gruneberg, H. 1943. *The Genetics of the Mouse.* London. Cambridge University Press.

Gubb, D. 1985. Domains, compartments and determinative switches in *Drosophila* development. *BioEssays* 2:27–31.

Hadorn, E. 1965. Problems of determination and transdetermination. *Brookhaven Symp. in Biol.* 18:148–161.

Hall, B. K. 1983. Epigenetic control in development and evolution. In B.C. Goodwin, N. Holder and C.C. Wylie, eds. *Development and Evolution*, pp. 353–379. Cambridge: Cambridge University Press.

Hall, W. G. and C. L. Williams. 1983. Suckling isn't feeding, or is it? A search for developmental continuities. *Advances in the Study of Behavior* 13:219–254.

Hennig, W. 1966. *Phylogenetic Systematics*. Urbana: University of Illinois Press.

Hinton, H. E. 1963. The origin and function of the pupal stage. *Proc. Roy. ent. Soc. Lond. A* 38:77–85.

Hoogland, R. D. 1951. On the fixing-mechanism in the spines of *Gasterosteus aculeatus* L. *Ned. Akad. Wet. Ser. C* 54:171–180.

Horder, T. J. 1981. On not throwing the baby out with the bath water. In G. G. E. Scudder and J. L. Reveal, eds. *Evolution Today: Proceedings of the Second International Congress of Systematic and Evolutionary Biology*, pp. 163–180. Pittsburgh: Hunt Institute for Botanical Documentation, Carnegie-Mellon University.

Horder, T. J. 1983. Embryological bases of evolution. In B.C. Goodwin, N. Holder and C.C. Wylie, eds. *Development and Evolution*, pp. 315–352. Cambridge: Cambridge University Press.

Hubbs, C. L. 1944. Concepts of homology and analogy. *Am. Nat.* 78:289–307.

Kalter, H. 1980. A compendium of the genetically induced congenital malformations of the house mouse. *Teratology* 21:397–429.

Kauffman, S. A. 1973. Control circuits for determination and transdetermination. *Science* 18:310–318.

Kauffman, S. A. 1983. Developmental constraints: internal factors in evolution. In B.C. Goodwin, N. Holder and C.C. Wylie, eds. *Development and Evolution*, pp. 195–225. Cambridge, U. K.: Cambridge U. Press.

Lawrence, P.A. 1981. The cellular basis of segmentation in insects. *Cell* 26:3–10.

Lewin, R. 1984. Why is development so illogical? *Science* 224:1327–1329.

Lindenmayer, A. 1982. Developmental algorithms: Lineage vs. interactive control mechanisms. In S. Subtelny and P. B. Green, eds. *Evolution and Development*, Berlin: Springer Verlag.

Mayr, E. 1982. *The Growth of Biological Thought*. Cambridge: The Belknap press of Harvard University Press.

Moment, G. B. 1945. The relationship between serial and special homology and organic similarities. *Am.Nat.* 79:445–455.

Nelson, G. 1978. Ontogeny, phylogeny, paleontology, and the biogenetic law. *Syst. Zool.* 27:324–345.

Nelson, G. 1985. Outgroups and ontogeny. *Cladistics* 1:29–46.

Nieuwkoop, P. D. and L. A. Sutasurya. 1983. Some problems in the development and evolution of the chordates. In B.C. Goodwin, N. Holder and C.C. Wylie, eds. *Development and Evolution*, pp. 123–135. Cambridge: Cambridge University Press.
Nijhout, H. F., G. Wray, C. Kremen, and C. Teragawa. 1987. Ontogeny, phylogeny, and the evolution of form: an algorithmic approach. *Syst. Zool.* 35:445–457.
Noden, D. M. 1982. Patterns and organization of craniofacial skeletogenic and myogenic mesenchyme: a perspective. In A. D. Dixon and B. G. Sarnat eds. *Factors and Mechanisms Influencing Bone Growth*, pp. 167–203. New York: Alan R. Liss.
Northcutt, G. and C. Gans. 1983. The genesis of neural crest and epidermal placodes: a reinterpretation of vertebrate origins. *Quart. Rev. Biol.* 58:1–28.
Oppenheimer, J. 1940. The non-specificity of the germ layers. *Quart. Rev. Biol.* 15:1–26.
Oster, G. and P. Alberch. 1982. Evolution and bifurcation of developmental systems. *Evolution* 36:444–459.
Owen, R., 1848. *On the Archetype and Homologies of the Vertebrate Skeleton.* London: R. and J.E. Taylor.
Patterson, C. 1982. Morphological characters and homology. In K.A. Joysey & A.E. Friday, eds. *Problems of Phylogenetic Reconstruction*, pp. 21–74. London: Academic Press.
Raff, R. A. and T. C. Kaufman. 1983. *Embryos, Genes, and Evolution.* New York: Macmillan.
Remane, A. 1952. *Die Grundlagen des naturlichen Systems der vergleichenden Anatomie und der Phylogenetik.* Leipzig: Geest & Portig.
Riedl, R. 1977. A systems-analytical approach to macroevolutionary phenomena. *Quart. Rev. Biol.* 52:351–370.
Riedl, R. 1978. *Order in Living Organisms: a Systems Analysis of Evolution.* New York: Wiley.
Riedl, R. 1982. A dialectic approach to epigenetics and macroevolution. In V.J.A. Novak and J. Mlikovsky eds. *Evolution and Environment*, pp. 41–50, Prague: Academid.
Roth, V.L. 1982. *Dwarf Mammoths from the Santa Barbara, California Channel Islands: Size, Shape, Development and Evolution.* Ph.D. thesis, Yale University.
Roth, V.L. 1984. On homology. *Biol. J. Linn. Soc.* 22:13–29.
Sander, K. 1983. The evolution of patterning mechanisms: gleanings from insect embryogenesis and spermatogenesis. In B.C. Goodwin, N. Holder and C.C. Wylie, eds. *Development and Evolution*, pp. 137–160. Cambridge: Cambridge University Press.
Sattler, R. 1984. Homology: A continuing challenge. *Syst. Bot.* 9:382–394.
Seigel, J. A. 1982. Median fin-spine locking in the ponyfishes [Perciformes:Leiognathidael]. *Copeia* 1982:202–205.
Simpson, G.G. 1975. Recent advances in methods of phylogenetic inference. In

W.P. Luckett & F.S. Szalay, eds. *Phylogeny of the Primates*, pp. 3–19. New York, Plenum Press.

Sober, E. 1984. Common cause explanation. *Philosophy of Science* 51:212–241.

Sonneborn, T.M. 1970. Gene action in development. *Proc. Roy. Soc. Lond. B* 176:347–366.

Stevens, P.F. 1984. Homology and phylogeny: morphology and systematics. *Syst. Bot.* 9:395–409.

Sulston, J. E., D. G. Albertson, and J. N. Thomson. 1980. The *Caenorhabditis elegans* male: Postembryonic development of nongonadal structures. *Devel. Biol.* 78:542–576.

Tomlinson, B. 1984a. Homology in modular organisms—concepts and consequences. Introduction. *Syst. Bot.* 9:373.

Tomlinson, B. 1984b. Homology: an empirical view. *Syst. Bot.* 9:374–381.

Van Valen, L. 1982. Homology and causes. *J. Morphol.* 173:305–312.

Whittle, J.R.S. 1983. Litany and creed in the genetic analysis of development. In B.C. Goodwin, N. Holder and C.C. Wylie, eds. *Development and Evolution*, pp. 59–74.

Wiley, E. O. 1981. *Phylogenetics: The Theory and Practice of Phylogenetic Systematics* New York: Wiley-Interscience.

Wolfram, S. 1984. Cellular automata as models of complexity. *Nature* 311:419–424.

2. Indirect and Direct Methods In Systematics

Peter H. Weston

Cladistic analysis in the most general sense involves the testing of different hypotheses of synapomorphy against one another to produce a hierarchical summary of the inferred relationships between taxa (see e.g., Eldredge and Cracraft 1980; Nelson and Platnick 1981; Wiley 1981; Farris 1983). Whether one takes the narrower view (as I do) that cladistics is a fundamentally phylogenetic method (e.g., Eldredge and Cracraft 1980; Wiley 1981) or alternatively that phylogeny is an explanation for the natural order revealed by cladistic analysis (e.g., Nelson and Platnick 1981) is not substantially important. In either case the approach is basically the methodological descendant of Hennig's Phylogenetic Systematics (1966) and ultimately of cladistic principles implicit in earlier evolutionary and pre-evolutionary systematics. All of these explicitly cladistic methods involve three main steps:

(i) the formulation of hypothetical, unpolarized, character transformation series on the basis of structural and/or developmental and/or functional correspondences;

(ii) the "polarization" of these transformation series, producing hypothetical character phylogenies in which the relatively apomorphous (less general, advanced, derived) characters (or character states) are derived from the relatively plesiomorphous (more general, primitive, ancestral);

(iii) the construction of a cladogram that combines the information provided by the character phylogenies using one or more criteria (e.g., parsimony).

Step (i) has aroused relatively little controversy among both proponents and critics of cladism. It seems to be generally agreed that the formulation of hypothetical structural correspondences between different organisms and taxa underpins any approach to comparative biology (see, e.g., Mayr 1969; Sneath and Sokal 1973; Wiley 1981; Sattler 1984).

Step (iii) has been the subject of some contention and a number of different criteria have been proposed, often allied with various assumptions concerning evolutionary processes (see, e.g., Felsenstein 1982). However, a strong, and to my mind convincing, case has been made by Farris (1982, 1983) for parsimony being necessary and sufficient for cladogram construction.

Step (ii), however, continues to generate controversy, both among cladists (see, e.g., Brooks and Wiley 1985 and Kluge 1985 vs. Nelson 1985) and critics of cladism (e.g., Sneath and Sokal 1973). Early criticisms of Hennig (1966) by pheneticists (e.g., Colless 1967, 1969; Sneath and Sokal 1973) asserted that this step fatally flawed the logical basis of cladistic analysis. Basically they claimed that none of Hennig's four criteria (correlation of series of transformations, geological precedence, chorological progression, and ontogenetic character precedence) for the polarization of character transformation series were both independent of preexisting phenetic analyses and logically valid. In particular they focused on what they regarded as Hennig's strongest criterion, that of the correlation of series of transformations. This includes Naef's "principle of systematic character precedence," (see Hennig 1966) now generally referred to as the method of outgroup comparison (e.g., Maddison et al. 1984). This method was seen as leading to an infinite regress because it relies on a preexisting higher-level phylogeny to serve as a framework within which outgroup comparisons can be made. The higher level phylogeny must also be based initially on some tentative hypotheses of synapomorphy which may also have been formulated using the outgroup method. Clearly, the appeal to ever higher-level phylogenies cannot go on forever and a method independent of any preexisting phylogenetic hypotheses must be invoked ultimately to validate outgroup comparisons. None of Hennig's three other criteria were seen to be adequate to fill this gap. Those of geological character precedence and chorological progression were claimed by Sneath and Sokal (1973) also to rely on a preexisting taxonomic analysis. The essence of this point of view was repeated by cladists such as Nelson (1973, 1974) who also criticized these criteria as being based on untestable or unreasonable assumptions. Hennig's remaining criterion, that of ontogenetic character precedence, that is, the idea that ontogeny recapitulates phylogeny was dismissed by Sneath and Sokal (1973:43) because "it is now realized that the many exceptions from this law can lead to serious errors of interpretation." Nelson (1973) also rejected the law of recapitulation as simply being an untestable assumption.

Given the success of these arguments, the pheneticists' criticism would be devastating. Lundberg (1972), however, proposed a compromise solution to the problem. He argued that a phenetic classification could be taken as a reasonable first approximation to a higher-level cladistic hypothesis and that this could be used as a framework for outgroup comparisons. This approach would be unsatisfactory however, making cladistic analysis (and therefore much of evolutionary biology) logically dependent upon phenetic taxonomy and its attendant assumption of homogeneity of evolutionary rates (see Farris 1971). Moreover, it is highly debatable whether existing higher-level classifications are phenetic anyway (see, e.g., Nelson and Platnick 1981, Mayr 1982, Stevens 1984). Certainly the bases of higher classifications were established long before phenetics appeared in a formal guise.

Nelson (1973) in a brief reply to Lundberg proposed a solution to this problem in his discussion of direct and indirect techniques for establishing character phylogenies. According to Nelson (1973:87) "The techniques available for estimation of ancestral conditions may be divided into (1) indirect arguments, involving consideration of species, or groups, other than those of immediate concern, and (2) direct arguments, not involving other species or groups." Outgroup comparison is the only indirect argument discussed by Nelson. The only direct technique that he considered to be valid involves the study of ontogenetic character transformations. Nelson showed that the logic of both techniques is based on parsimony—that is, they yield character phylogenies that require fewer postulated character transformations than alternative possibilities. An important aspect of this analysis was that it suggested a previously unrecognized unity and simplicity of method within cladistic analysis. Parsimony was shown to be a necessary and sufficient criterion for both cladogram construction (Farris et al. 1970) and the study of character phylogeny. Cladistic analysis could be reduced to a method in which the distribution of homologies is interpreted parsimoniously (cf. Patterson 1981:448).

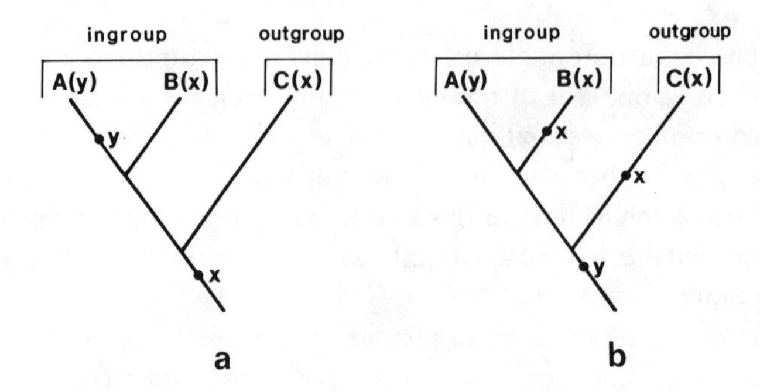

Figure 2.1. The indirect method and parsimony. Application of the indirect method (a), results in a character phylogeny one step shorter than the alternative (b). In this and all following diagrams in this chapter, a spot represents the gain of a character while an open circle represents the loss of a character.

The aim of this chapter is to reexamine the role of parsimony in Nelson's analysis. From this perspective I hope to demonstrate that his ontogenetic method is a special case of a more general direct method for polarizing character transformation series. I will also examine the relationship between Nelson's direct and indirect methods. Finally I will discuss some criticisms of Nelson's use of the parsimony criterion. First, however, I will deal with the indirect method, because it is relatively unproblematical.

THE INDIRECT METHOD

The logic of Nelson's indirect method is illustrated by the hypothetical example in figure 2.1, in which the cladistic structure represents a preexisting higher level phylogeny. The adult of taxon *A* possesses character *y* while those of taxa *B* and *C* possess its homologue, character *x*. The aim of the analysis is to determine which character was possessed by the most recent common ancestor of *A* and *B*. By applying the method of outgroup comparison (fig. 2.1a) a character phylogeny is derived that requires only two postulated evolutionary transformations, the appearance of *x* and its transformation into *y*. The alternative (fig. 2.1b) requires three steps and therefore is rejected on the grounds of parsimony.

Maddison et al. (1984) provide the most recent and detailed analysis of outgroup comparison in relation to parsimony. While implicitly endorsing the logic of Nelson's analysis they show that examination of only the sister group does not necessarily yield a uniquely most parsimonious solution. The next most closely related clade also must be examined to achieve this. In other words, a paraphyletic "doublet" of sister clades must agree if they are to yield a decisive parsimony argument. If the two clades of the doublet contradict one another however, further outgroup clades must be examined in order to apply other outgroup "rules." Maddison et al. discuss the effects of outgroup variability and the need for extra cladistic resolution within such outgoups before comparisons may be made. An important conclusion flowing from this analysis is that quite a highly resolved higher-level phylogeny is required before the indirect method can be applied rigorously. This further emphasises the need for a logically defensible direct method of character analysis.

THE DIRECT METHOD

Nelson summarized his direct method in the form of a rule, a reformulated biogenetic law, as follows: "Given an ontogenetic character transformation, from a character observed to be more general to a character observed to be less general, the more general character is primitive and the less general advanced" (Nelson 1978:327).

In the sense in which it is used here, "more general" does not simply mean "more common," although the more general character is in fact more common than the less general. What is important is that the more general character is possessed by all of the taxa that possess the less general character and also by some that do not. I will term such a relationship an unequivocal relationship of generality. This is illustrated in figure 2.2a in which the length of each bar represents the number of species possessing each character. The overlap in position of each bar indicates the "overlap" of the characters in taxa that exhibit both in ontogeny. This may be contrasted with the equivocal relationship of generality illustrated in figure 2.2b. Note that in the second example character x is more common than character y but is not more general than it.

The logic behind Nelson's direct method is shown in figure 2.3, a hypothetical example in which character y is a terminal ontogenetic stage developed from its homologue, character x. Given that both x and y must have been gained at some stage (i.e., that all characters are syna-

Figure 2.2. Generality relationship between homologous characters. (a) An unequivocal relationship; *y* is less general than *x*; (b) an equivocal relationship: neither character is less general than the other, even though *x* is more common than *y*.

pomorphies at some level) then the hypothesis that *y* is apomorphous relative to *x* (fig 2.3a) is more parsimonious by one step than its alternative (fig 2.3b) and is therefore logically preferable.

It must be emphasized at this point that although the conclusions derived using Nelson's direct method are identical to those drawn from "recapitulationary" arguments, it does not rely on any assumption of ontogeny recapitulating phylogeny. Indeed Nelson's method is a technique for detecting whether an ontogenetic sequence recapitulates phylogeny; recapitulation is a conclusion rather than a premise of the argument. Not all ontogenetic sequences logically need to be recapitulations of phylogeny, since the method is based on comparative generality of characters rather than on any assumed analogy between phylogenetic and ontogenetic transformations. This point is best illustrated using two hypothetical examples. The first is a trivial case in which one class of organisms possesses character *x* as a terminal ontogenetic stage while in all others possessing character *x* it is transformed into *y* and then into *z*

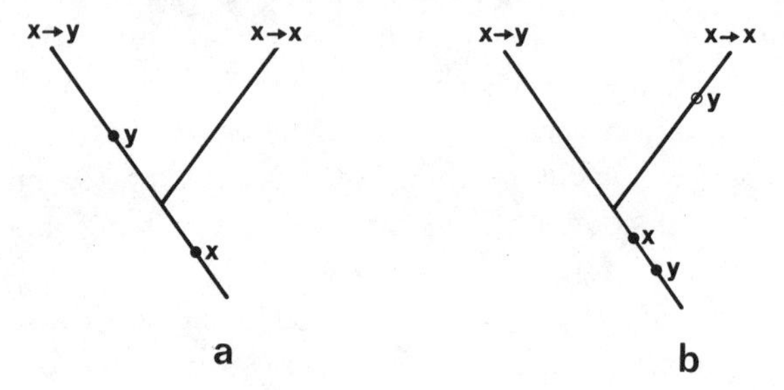

Figure 2.3. The direct method and parsimony. Application of the direct method (a) results in a character phylogeny one step shorter than the alternative (b).

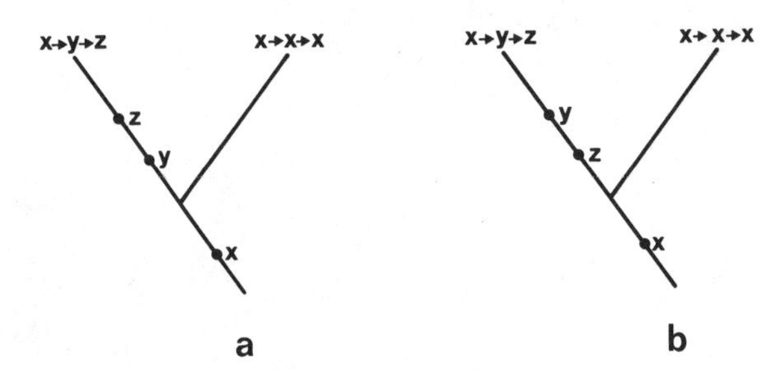

Figure 2.4. Hypothetical example showing the conceptual difference between relative generality and ontogenetic sequence. Even though *y* precedes *z* in ontogeny, the two characters are equally general and alternative character phylogenies (a & b) are equally parsimonious.

(fig. 2.4). Under a recapitulationary argument *y* would be regarded as plesiomorphous relative to *z* but apomorphous relative to *x*. The application of Nelson's method, however, would find *y* and *z* to be equally general and therefore would conclude that neither is more apomorphous than the other. The second example involves characters showing an equivocal relationship of generality as in figure 2.2b. In this case the alternative character phylogenies (fig. 2.5) are equally parsimonious, and neither can be preferred on grounds of parsimony alone despite the existence of an ontogenetic transformation from *x* to *y* in some species.

Given that the relative generality of homologous characters, and therefore parsimony, forms the crux of Nelson's direct method, it is reasonable to ask whether the order of appearance of characters in ontogeny has

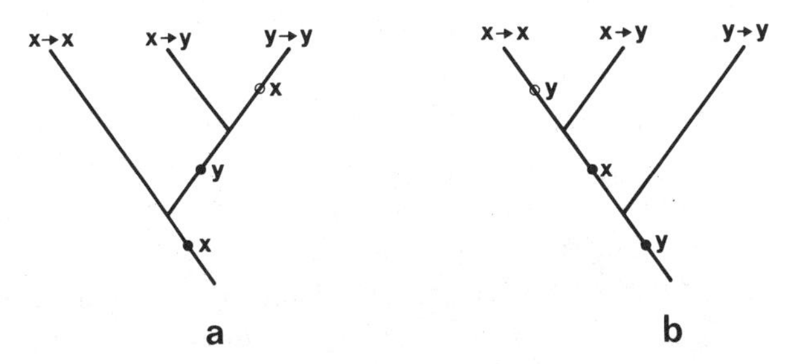

Figure 2.5. Given an unequivocal relationship of generality between characters *x* and *y*, the alternative character phylogenies (a & b) are equally parsimonious.

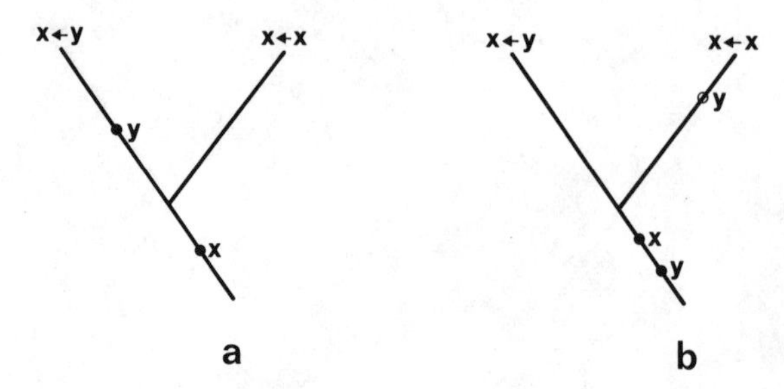

Figure 2.6. Parsimony and direction of ontogenetic sequence. By reversing the arrows in figure 2.3, this figure is produced. The parsimonious representation of character phylogeny, however, remains unchallenged.

any special significance as Nelson's biogenetic law suggests (cf. de Queiroz 1985). For example, by reversing the arrow in figure 2.3 to create figure 2.6 we do not change the parsimony argument at all. Character *y* is still less general than *x* and can be postulated as relatively apomorphous on grounds of parsimony. On the basis of this hypothetical example we could conclude that the exact transformational sequence or pathway is irrelevant to the operation of the "direct method." So long as more than one homologous character can be present in the ontogeny of an organism and an unequivocal relationship of generality exists, Nelson's direct method can be applied.

This conclusion, however, would ignore an important feature of the particular examples that Nelson used to illustrate his argument. These are all classic examples of vertebrate ontogeny and are typified by the following case. The embryos of all vertebrates develop pharyngeal gill slits at an early stage of development and in some species (fishes, some amphibians) these remain open through to the adult stage. In the others the gill slits close before the adult stage is reached. Here, open gill slits are equivalent to character *x* in figure 2.3 while closed gill slits are equivalent to character *y*. This transformation occurs uniquely at a certain stage in the development of each individual so that the character "open gill slits" is lost irreversably when the character "gill slits closed" is gained. In this and other similar examples from vertebrate development, each ontogenetic stage is a modification of, or is induced by, a preexisting ontogenetic stage and so is developmentally dependent upon the successful development of its precursor. In Løvtrup's (1978) terminology these are "epigenetic characters." If the developmental dependence of characters remains unaltered by subsequent evolution then the sequence of appearance of particular characters cannot be changed. For example, given the epigenetic sequence *a→b→c→d,* novel characters could be added or deleted terminally (*a→b→c→d→e* or *a→b→c*) or changed subterminally (*a→b→e→f*). However, novel characters could not be inserted (*a→b→e→c→d*) or deleted (*a→c→d*) subterminally without a fundamental change in the developmental interactions between characters. Similarly, the sequence could not be "shuffled" (*a→d→b→c*). Under this developmental model, only terminal additions create information that can be analysed using Nelson's biogenetic law. Subterminal changes result in the loss of useful information, necessitating the use of outgroup comparison for fully resolving character transformation series. Terminal deletions (neotenic reversals) are misleading in that they are a source of homoplasy that must be detected by comparative analysis of other characters. Nevertheless, since subterminal deletions and insertions cannot occur, we can assume that the ontogenetic sequence of characters obeys von Baer's law and proceeds from more general to less general—i.e., from primitive to advanced. The sequence of appearance of characters in ontogeny therefore recapitulates character phylogeny.

According to Alberch (1985), however, it would be naïve to assume that all ontogenetic sequences are this inflexible. He provides examples of ontogenetic sequences in which epigenetic interactions themselves must have changed as a result of evolution. Similarly, Rosen (1984) and Kluge (1985) give examples of partially reversible ontogenetic sequences in which mature tissues and structures have the potential to "dedifferentiate." Clearly, the order of appearance of characters in ontogeny will be reliable indicators of character phylogeny only if epigenetic constraints are strong and conserved during the course of evolution. This caveat leads us back to the notion that relative generality of homologous characters, not their sequence in ontogeny, is the primary criterion of Nelson's direct method. What then is the significance of ontogeny in the logic of Nelson's direct method? As Nelson himself has noted, "the mode of development itself is the most important criterion of homology" (Nelson 1978: 335). It may not be a necessary or sufficient criterion for formulating homologies but it is a very strong one nevertheless (cf. Kaplan 1984). Alberch's point that apparently homologous characters may develop through quite incomparable ontogenies in different species does not diminish the strength of this argument. The observation that character *x* transforms into *y* during ontogeny is still a very good reason to postulate them as homologous.

GENERALIZING NELSON'S DIRECT METHOD

Nelson's version of the biogenetic law precisely specifies the circumstances under which it is applicable: it defines a single type of ontogenetic transformation that may be used for directly inferring character phylogenies. As I have shown above, however, it is possible to conceive of hypothetical examples that do not conform to this model but which may still be analysed using the basic parsimony argument of Nelson's method. In this section I will give some corresponding real examples. I will also examine a number of other real examples that are not interpretable under Nelson's biogenetic law but which may be analyzed using a more broadly framed direct method of character analysis.

Subterminal insertions and deletions

Rosen (1982:76–79; 1984:79–85) furnishes an example involving two ontogenetic sequences (fig. 2.7), one partly comparable to figure 2.5 (Rosen's characters *A*1–3) and the other comparable to figure 2.6 (Rosen's characters *B*1–3). Neither sequence can be analyzed under Nelson's biogenetic law because in neither case does the sequence proceed from more general to less general. Clearly such cases must involve either subterminal insertions or deletions and so the epigenetic interactions between the characters must have changed during their evolution. The most parsimonious explanations of character phylogeny are shown in figure 2.7. Neither explanation involves recapitulation but rather the insertion of novel characters earlier in the developmental sequence. Interestingly, Rosen analyzes both sequences as recapitulations. In so doing he is treating those characters that appear earlier in ontogenetic sequences as more general than those that appear later (assuming von Baer's law to be unreservedly true) despite the existence of data to the contrary. A corollary of this approach is the assumption that subterminal stages can be deleted from, but not inserted into, ontogenetic sequences. As Nelson (1985) points out, this approach is one of "constrained parsimony." That is, given one or more constraints, the analysis provides the most parsimonious solution. In this respect Rosen's use of von Baer's law is com-

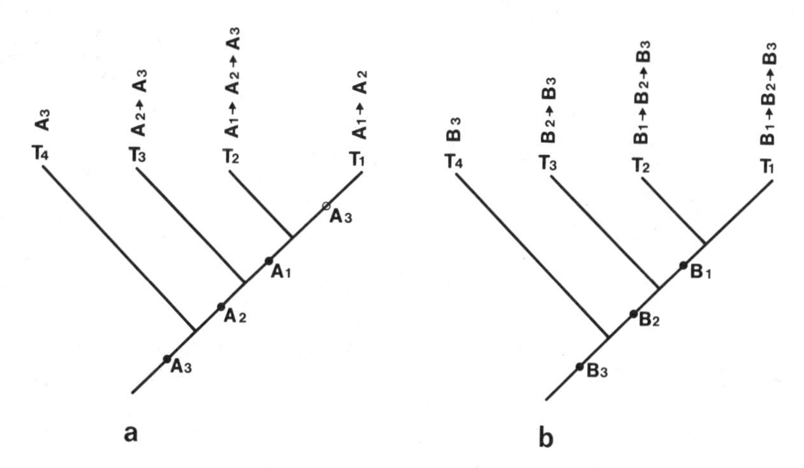

Figure 2.7. Parsimonious character phylogenies for ontogenetic sequences discussed by Rosen (1982, 1984). T1–T4 represent taxa and the sequences illustrated above them represent ontogenetic transformations.

parable to the normal cladistic constraint that relationships be hierarchical, not reticulate. Other constrained parsimony methods include the Camin-Sokal and Dollo parsimony procedures reviewed by Felsenstein (1982). However, since constrained parsimony procedures invoke additional assumptions over and above the conventions of parsimony and hierarchy, it is reasonable to ask whether these assumptions are justified. Rosen's justification appears to be that ontogenetic transformations provide the only solid empirical basis for cladistic character transformations and that phylogeny is simply "ontogeny writ large" (cf. Patterson 1982). However, this approach implies the axiomatic acceptance of ontogenetic recapitulation that I believe Nelson's direct parsimony argument was designed to avoid.

If we are prepared to weight subterminal deletions and insertions equally we can broaden Nelson's direct method, removing any reference to the order of characters in a sequence.

The ontogeny of serially homologous characters

All of the examples that I have considered so far involve the direct, sequential transformation of one structure into another during development. The ontogeny of serially homologous characters, however, in-

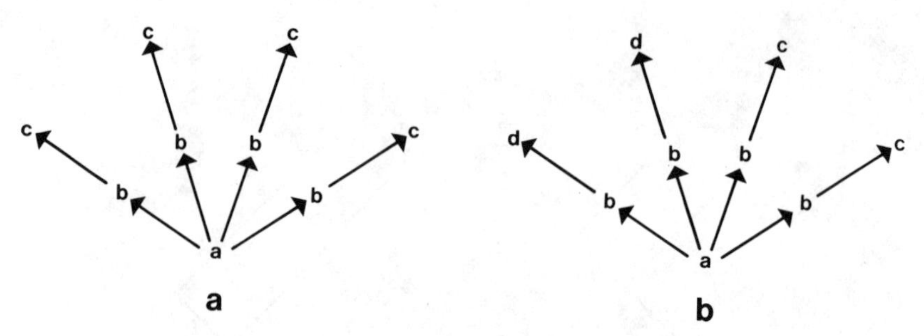

Figure 2.8. Serially homologous characters represented as branching ontogenetic pathways. Some types of branch (*b*→*d* are less general than the other (*b*→*c*).

volves the development of a number of adult structures from a single type of ontogenetic precursor and ultimately from a single primordial structure or tissue. This can be represented as a branching ontogenetic tree, as in figure 2.8a. Since there may be little or no epigenetic interaction between the different "branches," it would be possible for some to transform phylogenetically into novel structures while others in the same organism remain unaltered (fig. 2.8b). This would result in a relationship of generality between characters *c* and *d* that is equivalent to that between characters *x* and *y* in figure 2.3 despite the absence of any direct ontogenetic transformation from *c* to *d*.

Consider an example from higher plants: the ontogenetic development of leaves in *Acacia* (Fabaceae: Mimosoideae). Some *Acacia* species produce true, compound leaves throughout their ontogeny. Others, however, produce true, compound leaves only as juvenile foliage and these are replaced by simple phyllodes early in the life of the plant (Cambage 1915–28; Kaplan 1975; Vassal 1979). Usually a range of intermediate, compound leaf/phyllode organs are produced at this transition. Since compound leaves are produced by all species that produce phyllodes and also by some that do not, compound leaves are more general than phyllodes and we may postulate them to be more primitive than phyllodes following the parsimony argument of Nelson's direct method.

Phyllodes, however, are not ontogenetically transformed versions of compound leaves. Starting at an early stage in leaf organogeny, each leaf type is recognizably different from the other (Kaplan 1975, 1980, 1984). The latest organogenetic precursor common to all leaf types is the undifferentiated leaf primordium. Pinnate and bipinnate compound leaves share a later stage in common, an embryonic compound leaf with primordial pinnae.

Figure 2.9 is a diagrammatic summary of the early shoot ontogeny of a phyllodineous *Acacia* plant. The only transformation that occurs in the ontogeny of the entire plant is a transformation in the behavior of the apical meristem. Any "transformation" in leaf type is an illusion produced by the intergrading series of discrete leaf types. The apical meristem, however, does not undergo an irreversible ontogenetic transformation as do a mammal's embryonic gill slits. A meristem is in many respects a perpetually embryonic tissue which is potentially capable of producing any of the organs in its developmental "repertoire." The organs that it does produce are largely determined by the control of hormones and growth substances that in turn are regulated by genetic and/or environmental factors. It is hardly surprising then that reversals in meristem behavior occur, producing so-called reversion shoots either as coppice shoots or spontaneously on adult shoots in species such as *A. melanoxylon* (Fletcher 1920; Kaplan 1975). It therefore seems doubtful that a phyllode-producing meristem is developmentally dependent on the prior development of a compound-leaf-producing meristem. Using the terminology of Alberch (1985), the ontogenetic sequence from compound leaves to phyllodes should be regarded as a temporal rather than causal sequence. Consequently, leaf ontogeny in *Acacia* can be summarized as a branching diagram (fig. 2.10) with "meristem" at its root. This does not mean that leaves are homologous with the apical meristem but that they develop from part of the apical meristem.

It seems reasonable to conclude that a novel leaf type could be inserted phylogenetically anywhere in the ontogenetic sequence of *Acacia* without affecting the development of the already existing leaf types. Graphically, this can be represented on figure 2.10 by adding in a new branch onto "1" (leaf primordium). If this conclusion is correct, then the order of appearance of particular leaf types in heteroblastic sequences would be irrelevant in applying Nelson's direct method of character analysis. More generally, any transformational sequence involving meristem behavior would seem to be unreliable as a direct indication of character phylogeny. What is informative in such cases is the relationship of generality (if it is unequivocal) between the different serially homologous characters. The fact that the ontogenetic sequence runs from more general to less general

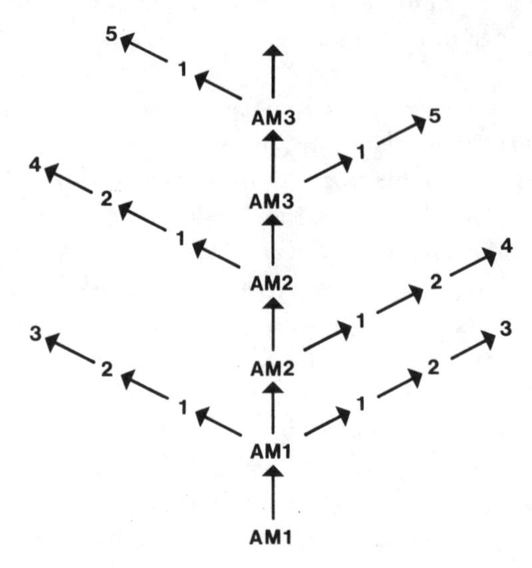

Figure 2.9. Diagramatic summary of early shoot ontogeny in a phyllodineous *Acacia* species. Symbols: AM1–AM3, apical meristem at different stages in its ontogeny; 1, leaf primordium; 2, compound leaf primordium; 3, pinnate compound leaf, 4, bipinnate compound leaf; 5, phyllode.

in *Acacia*, thus recapitulating character phylogeny, might be due to natural selection rather than inflexible developmental constraints (c.f. Stebbins 1974:113–114).

In many groups showing heteroblastic leaf sequences, the juvenile leaf

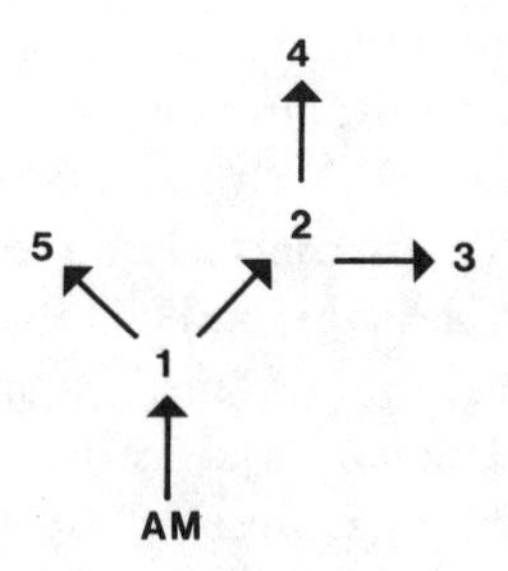

Figure 2.10. Branching ontogenetic pathway summarizing leaf organogeny in a phyllodineous *Acacia* species. Symbols: AM, apical meristem; others as in fig. 2.9.

characters are less general than those of the adults. For example, several New Zealand species of *Pseudopanax* (Araliaceae), the "lancewoods," produce spectacularly distinctive juvenile leaves in which the leaflets are rigid, extremely long and narrow, and are deflexed (Philipson 1965, and references therein; Salmon 1980). The adult leaflets of these species, however, resemble both the adult and juvenile leaflets of other species in being softer, much shorter, broader, and upright-spreading. On the basis of a generalized direct method, the juvenile characters of the "lancewoods" listed above could be postulated as synapomorphous for those species.

Cell and tissue types may also exhibit branching ontogenetic sequences amenable to analysis under a generalized direct method. For example, one of the synapomorphies characterizing vascular plants is the development of xylem tracheary elements. All vascular plants produce at least some tracheids as tracheary elements during their ontogeny while only some also produce vessels (see, e.g., Cutter 1969). Vessels are directly observed to be less general than tracheids and so it can be concluded that they are relatively apomorphous. However, neither type of tracheary element develops directly from the other in ontogeny. Instead both types diverge during their development from the same type of provascular cell.

Organogenetic sequences

An important distinction I alluded to above is that between ontogeny and organogeny in higher plants. Ontogeny usually refers to the development of the whole *genet,* the whole living plant, largely reflecting the development of the apical meristems. Organogeny refers to the development of individual organs from their primordia. Organogenetic sequences are repeated, often indefinitely, throughout ontogeny. Consequently, higher plants can be said to have a modular construction in contrast to the unitary construction of most animals (Tomlinson 1982). Whereas ontogenetic sequences of plants are likely to be developmentally and phylogenetically labile, organogenetic sequences seem to be comparable with epigenetic sequences in animals in their developmental inflexibility (see, e.g., Sachs 1982). It follows that von Baer's law should usually apply to organogenetic sequences and that Nelson's biogenetic law will be potentially applicable to them.

An example where Nelson's law is applicable is the shoot development of *Acacia verticillata* (Hara and Kaplan 1980; Kaplan 1984). The "phyllodes" of this species are mostly grouped into whorls or fascicles of 2 to 9. However, about 1 in 8 (1 in each whorl) of these is a true leaf, homologous with the leaves of other *Acacia* species. Each true phyllode is stipulate, possesses a basal nectary, and is innervated from a three-traced unilacunar node. The other "phyllodes" are exstipulate, lacking nectaries, and are innervated by a single trace without a leaf gap. The organogeny of each whorl proceeds as follows. The first primordium to form is that of the true leaf. Later the stipule primordia differentiate laterally from the primordial leaf base. The primordia of the exstipulate appendages then develop in a transverse sequence laterally from the stipule primordia in a collarlike extension of the leaf base. At this early stage in organogeny, the stipulate and exstipulate "phyllodes" are quite dissimilar in appearance but as they develop they converge morphologically until at maturity they are morphologically and anatomically almost indistinguishable. In other *Acacia* species the development of lateral primordia ceases after the formation of stipule primordia. Development therefore proceeds from the more general characters (true leaf plus stipules) to the less general (exstipulate appendages). Under the direct method exstipulate appendages may be postulated to be apomorphous relative to phyllodes, notwithstanding the indefinite repetition of this sequence from an early stage in ontogeny.

Biosynthetic pathways are a class of ontogenetic pathways that are analagous to organogenetic sequences in that they may be repeated indefinitely throughout ontogeny. Seaman and Funk's (1983) paper on sesquiterpene lactone biosynthesis in the Asteraceae provides a good example of data amenable to analysis using the direct method. Sesquiterpene lactone biosynthesis in the four species of *Tetragonotheca* involves the production of an epoxyangulate sidechain, represented in figure 2.11a by 5. In two species this pathway produces an additional novel step (6 in figure 2.11b-c). Each of these species possesses its own unique end products (7,8 in fig. 2.11b, 9 in fig. 2.11c). Application of the direct method yields a character phylogeny as shown in figure 2.11d. However, Seaman and Funk (1983:8) assert that:

> Ontogenetic considerations are not really applicable to sesquiterpene lactone data because plant chemistries do not display any reliable or predictable developmental sequence which corresponds to the plant's life cycle. Rather than observing a progressive, step-wise synthesis of a series of homologous compounds during the

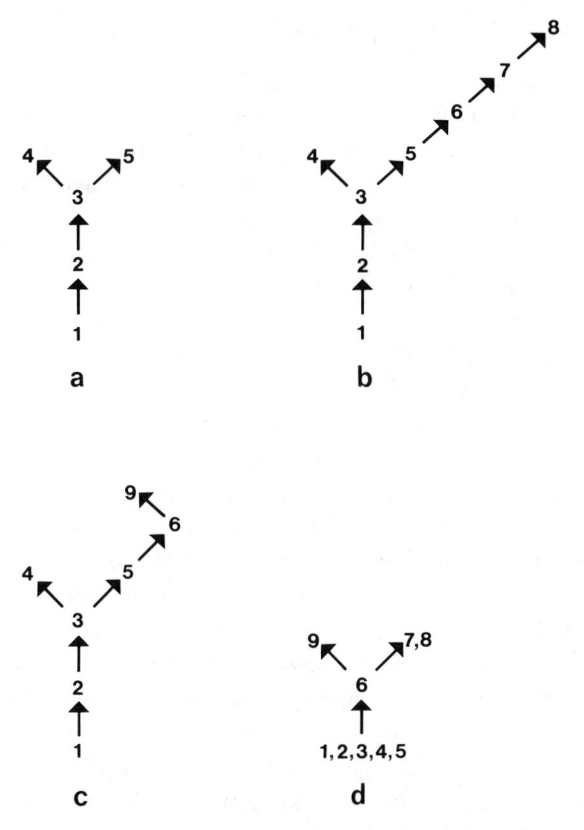

Figure 2.11. Biosynthetic pathways of the epoxyangelate sidechain of sesquiterpene lactones in *Tetragonotheca* (data from Seaman and Funk 1983). Numbers represent biosynthetic intermediates and end products. (a-c) represent biosynthetic pathways in different species. (d) represent the parsimonious character phylogeny derived using the direct method.

course of plant development, overall synthesis seems to be initiated when the sesquiterpene lactone-synthesising glandular tissues have differentiated.

This argument is incorrect since the relative generality of the characters involved is unequivocal. The fact that the ontogenetic pathway is not unique but occurs repeatedly within the life cycle in a particular tissue is irrelevant to the successful application of Nelson's direct parsimony argument.

Concluding remarks on a generalized direct method

The examples that I have discussed in this section indicate the profitability of broadening the applicability of Nelson's direct parsimony argument. Branching ontogenetic pathways that result in serially homologous characters are a class of ontogenetic relationships that have not been used generally in directly inferring the polarity of character transformation series. When they have been considered, it has been the temporal sequence of appearance of serial homologs, not their relative generality, that has been regarded as informative (e.g., Stebbins 1974). Since the ontogeny of higher plants largely reflects the development of serial homologs, this type of application of the direct method obviously has great potential in botanical cladistics. Similarly, organogenetic sequences in plants and biosynthetic sequences have been largely ignored by cladistic systematists as a source of direct information on character phylogeny. Biosynthetic pathways seem particularly significant since they introduce an ontogenetic dimension to organisms that lack ontogeny in the usual sense (cf. Kluge 1985). The use of these types of ontogenetic pathways would increase the scope of application of Nelson's direct method greatly. An increase in scope is obviously needed as shown by several recent attempts to deal with very high level cladistic relationships using only the indirect method to polarize character transformation series. Lipscomb's (1985) interesting and useful analysis of eukaryote relationships is a prime example. She constructed a Wagner network linking major eukaryote groups, based on ultrastructural characters and to that extent her analysis was unproblematic. However, when it came to rooting the network, she was embarrassed by her inability to make meaningful outgroup comparisons with prokaryotes. In the end, she relied on the red algae as her outgroup because "although certainly eukaryotes, they are generally considered an unusually primitive group" (Lipscomb 1985:132). If this approach is not just an irrational appeal to authority then there must be some logical basis for regarding the eukaryotes as monophyletic and the Rhodophyceae as the sister group to the rest of the eukaryotes. Since outgroup comparisons are uninformative, some direct method for establishing tentative synapomorphies must have been used in coming to these conclusions. I suggest that a generalized direct method could conceivably be used in formulating character phylogenies of ultrastructures based on their biosynthesis.

In focusing emphasis on a directly observed relationship of generality between characters, the role of the sequence of characters in ontogeny is diminished. In fact, recapitulationary sequences are seen to be the consequence of a particular type of inflexible ontogenetic relationship between characters. Since that type of ontogenetic relationship is not relevant to some types of generality relationship and may itself be subject to evolutionary modification (Alberch 1985), any reference to ontogenetic sequence should be deleted from a generalized direct method for polarizing character transformation series. The following methodological rule is general enough to cover my examples.

Given a distribution of two homologous characters in which one, x, *is possessed by all of the species that also possess its homolog, character* y, *and by at least one other species that does not, then* y *may be postulated to be apomorphous relative to* x.

THE RELATIONSHIP BETWEEN DIRECT AND INDIRECT METHODS

Nelson (1978) examined in great detail the logical relationship between the direct and indirect methods. He treated each method as a separate hypothesis and then enumerated the different classes of potential "falsifiers" of each hypothesis. Since the direct method is protected from falsification by fewer ad hoc explanations than the indirect method, Nelson concluded that it is of primary significance for cladistic analysis. This difference in "falsifiability" reflects the reliance of outgroup analysis on a higher level classification: anomalous results can be explained by the ad hoc hypothesis that the supposed outgroup is really an ingroup. This argument reduces to the observation that the direct method does not rely on any auxiliary theories whereas the indirect method does.

Nelson's treatment of methods as hypotheses, however, is an unusual approach in that methods are normally regarded as metaphysical constructs that cannot be falsified (see, e.g., Popper 1980). Since parsimony forms the basis of both direct and indirect methods, parsimony itself would be the concept falsified by a "refutation" of either method. Parsimony of course has never been regarded by cladists as open to falsification. It is a methodological rule or convention (i.e., a metaphysical concept) used to interpret our observations. Whether or not Nelson's

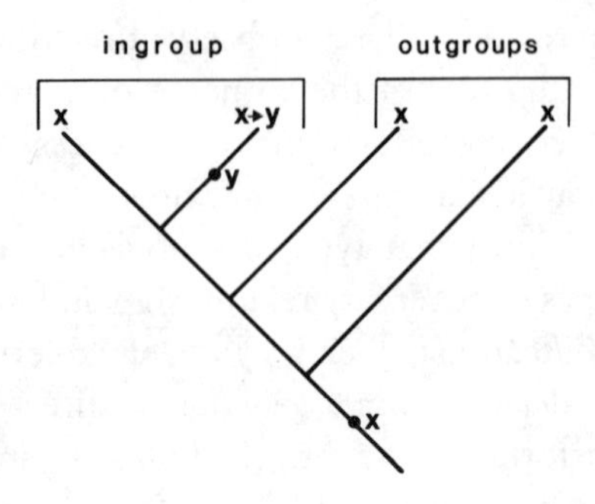

Figure 2.12. Agreement of direct and indirect methods.

"biogenetic law" really is a natural law (see Voorzanger and van der Steen 1982; Kluge 1985 vs. Nelson 1985) is thus irrelevant to the application of the direct method in cladistic analysis.

Since both the indirect and direct methods are applications of parsimony, it is possible to compare their performance directly in dealing with particular problems. While results of the two methods will be congruent in the absence of homoplasy (fig. 2.12), they may also disagree (fig. 2.13), or one or other may yield equivocal conclusions as a result of homoplasy (figs. 2.14–17). In the event of disagreement, which method, if any, is more reliable? Let us consider this question first under the assumption that the higher level classification we are using as a framework is infallible. Under such an assumption outgroup comparisons always indicate a more parsimonious solution than do conflicting conclusions derived using the direct method (fig. 2.13).

It would be useful if one type of character analysis could resolve problems for which the other method yields equivocal conclusions. Under the assumption made above, however, only outgroup comparison is capable of reliably resolving such problems (figs. 2.16, 17). The use of the direct method in this way is parsimonious for some character distributions (fig. 2.15) but not uniquely so for others (fig. 2.14). Under the assumption of an infallible higher level classification the indirect method should be preferred over the direct method. However, the only case in which this assumption seems defensible is the grouping of life as a clade. Making outgroup comparisons with non-life is equivalent to accepting that all characters must be synapomorphies at some level. This notion is inherent in the parsimony arguments behind both direct and indirect methods.

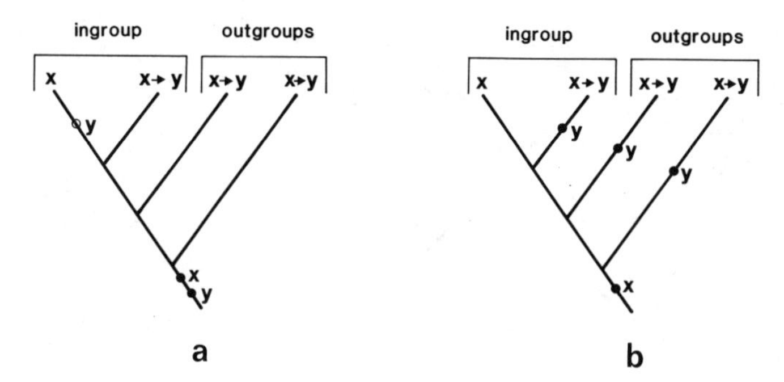

Figure 2.13. Disagreement of direct and indirect methods. The indirect method yields a character phylogeny that is more parsimonious than that introduced by the direct method, given the constraints of the higher-level phylogeny. (If the constraints of the higher-level phylogeny are ignored, then the opposite is true.)

Under the more reasonable assumption that our higher level classification is possibly wrong, resolution of conflict between the direct and indirect methods may be complicated. It involves reconsidering evidence from all available characters at the higher level in deriving a parsimonious solution. This in turn will require critical examination of the justification of higher level synapomorphies. Ultimately, some of that justification must stem from application of the direct method. Put another way, rejection of a particular application of the direct method in favor of the result of outgroup comparison will impose a parsimony "debt" that must be balanced by a larger parsimony "saving" in the interpretation of other characters.

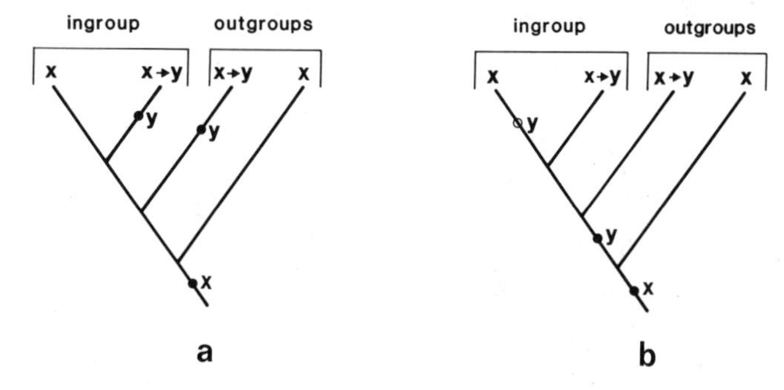

Figure 2.14. Equivocal indirect method and unequivocal direct method. Alternative character phylogenies are equally parsimonious.

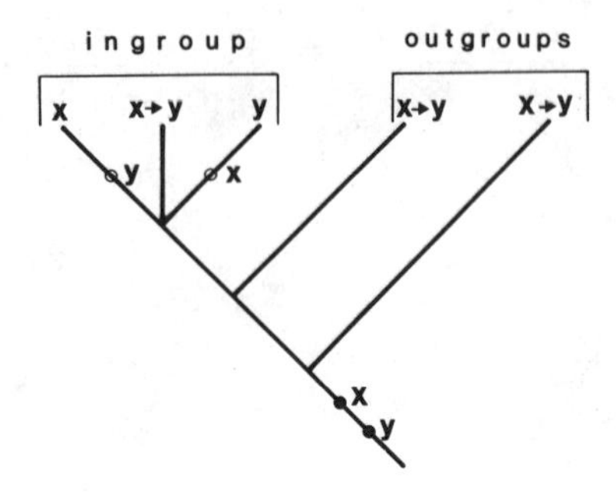

Figure 2.15. Equivocal indirect method and unequivocal direct method. Unique parsimonious character phylogeny is produced by the direct method.

The relationship between direct and indirect methods may be understood better if outgroup comparison is considered as a special case of the general use of parsimony in comparative analysis (c.f. Maddison et al. 1984). That is, outgroup comparison is simply a matter of extending the scope of parsimony analysis outside the group of immediate concern. The use of parsimony in comparative analysis involves the testing of different character phylogenies against one another as well as the polarization of undirected transformation series using those that are already known with confidence as points of reference. The scope of comparative parsimony analysis encompasses all organisms and all available characters.

By contrast, the scope of the direct method is limited in any particular

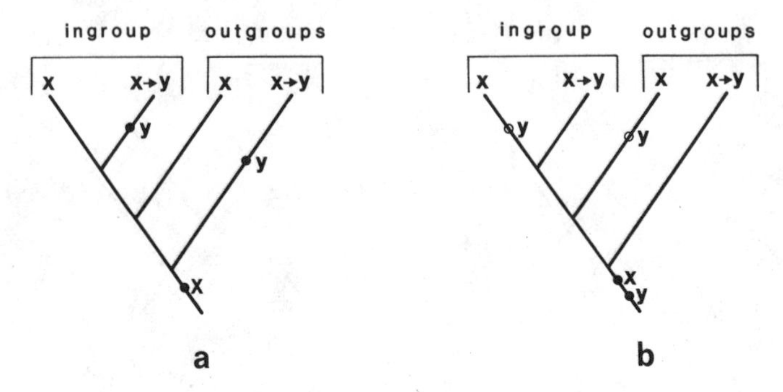

Figure 2.16. Equivocal direct method and unequivocal indirect method. Unique parsimonious character phylogeny is produced by the indirect method.

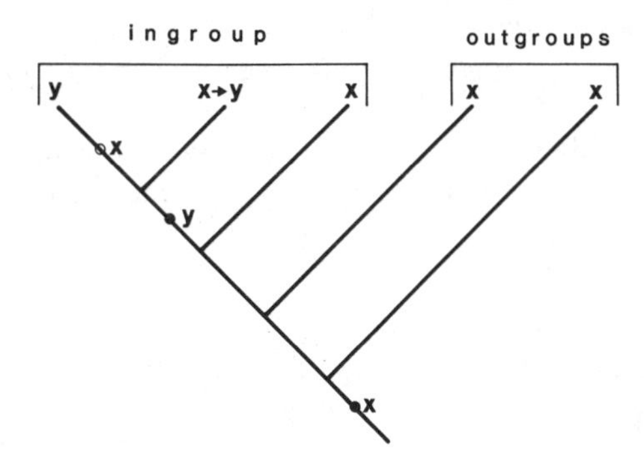

Figure 2.17. Equivocal direct method and unequivocal indirect method. Unique parsimonious character phylogeny is produced by the indirect method.

application to one set of homologous characters, and the organisms in which they occur. In relation to comparative analysis, it is a "hypothesis-generating" procedure, not a "hypothesis-testing" one: an equivocal relationship of generality does not refute any particular homology, but just renders the method inapplicable in that particular case. The direct method is used to produce tentative hypotheses of character phylogeny that may be tested against others using the comparative method.

It follows that the direct and indirect methods fulfill complementary roles in cladistic analysis. Tentative character phylogenies are required to give comparative analysis some fixed points of reference and these can be provided by the direct method. No character phylogenies, however, should be regarded as infallible, and it is the function of comparative parsimony analysis to test all character phylogenies, including those derived using the direct method, against one another in deriving a most parsimonious solution. Both methods are necessary, contrary to Brooks and Wiley's (1985:10) claim that "Direct observation of ontogeny as a method for polarising characters stands in relationship to outgroup comparisons also as a part to a whole." The structure of cladistic knowledge therefore, resembles Popper's view of science as a whole

> The empirical basis of objective knowledge has thus nothing "absolute" about it. Science does not rest upon solid bedrock. The bold structure of its theories rises, as it were, above a swamp. It is like a building erected on piles. The piles are driven down from above into the swamp, but not down to any natural or 'given' base; and if we stop driving the piles deeper, it is not because we have

reached firm ground. We simply stop when we are satisfied that they are firm enough to carry the structure, at least for the time being. (Popper 1980:111)

Comparative biology is like the building and applications of the direct method are like the piles. It would seem that the more "piles" there are supporting the "building," the better.

CRITICISMS OF NELSON'S DIRECT PARSIMONY ARGUMENT

Lundberg (1973) criticized Nelson's direct method on a basis that has been restated subsequently in various forms by Alberch (1985), Kluge (1985), and de Queiroz (1985). The gist of these arguments is that the characters in Nelson's example (see fig. 2.3) can be redefined so as to render the alternative character phylogenies equally parsimonious. Lundberg's reasoning can be paraphrased as follows: if the transformation $x \rightarrow y$ is regarded as a single character, rather than as two characters, then the primitive gain of $x \rightarrow y$ and the subsequent loss of y involves only two steps, not three as Nelson's argument has it. The primitive gain of x and the subsequent gain of y also involves two steps, so the alternatives are equally parsimonious. However, to my mind this is a peculiar way of viewing characters and their parsimonious representation. If characters are regarded as the parts of ontogenies that differentiate different organisms, then surely x and y can be regarded logically as different characters. Some species are characterized by the absence of both x and y (and therefore are irrelevant to the analysis), others by the presence of x and the absence of y, and yet others by the presence of both x and y. When characters are viewed this way Lundberg's argument collapses.

The essence of Lundberg's argument has been developed by de Queiroz (1985) in his definition of whole ontogenetic transformations as characters. De Queiroz objects to the use of "instantaneous morphologies" as characters because they are abstractions taken from real ontogenies or life cycles, and it is the phylogenetic relationships between real life cycles and not those between abstractions that we are interested in. His approach is a holistic one, as opposed to the more reductionary approach of analyzing the parts of ontogenies. Holism as a general approach, however, has been severely criticised (e.g., by Popper 1961:76–83) as unscientific: taken to an extreme, it implies the acceptance of the universe as an indivisible entity, no parts of which should ever be studied in isolation. De Queiroz, for example, notes that his approach, if taken to an extreme, would treat each organism as a single character and we would be left with no basis for comparative biology. He then retreats to a more reductionary approach:

The question then is "How large a segment of ontogeny constitutes a systematic character?" If systematic characters are defined as features of organisms that are used to determine the relationships among these organisms, then the answer to the question is "large enough to encompass variation that is potentially informative about the relationships among the organisms being studied." (de Queiroz 1985:293)

This definition of "character" closely resembles the one that I have adopted above in defending Nelson's parsimony argument against Lundberg's criticism. The definitions become practically identical if the clause "large enough to encompass variation" is replaced by "small enough to differentiate variation." In effect, de Queiroz is trying to invalidate Nelson's argument simply through arbitrary redefinition of a term.

Alberch (1985) notes that Nelson's direct method requires homologies to be established between embryonic and adult features, and that this requirement may be problematic. If accepted, this criticism has the same effect as that of Lundberg: alternative character phylogenies become equally parsimonious. It seems to me though that Alberch's criticism is relevant to a more general problem than the one he specifically addresses: the problem of suitable criteria for the recognition of homologies in systematics. This problem is distinct from that of the conceptual definition of homology, a problem resolved to my satisfaction by Patterson (1982) synonymizing homology with synapomorphy. Characters are postulated as homologous on the basis of similarity of position, construction, composition, ontogeny, etc. (see, e.g., Wiley 1981) and this is true whether the characters compared are from different stages of the same organism or "comparable" stages of different organisms. Characters thus homologized are never identical to one another and the homologies postulated may always be erroneous. In either case they are tested ultimately by comparative analysis.

The common theme running through the arguments outlined above is that operational definitions of "character" and "homology" can be adopted that invalidate Nelson's direct parsimony argument. This of course does not imply that these alternative definitions necessarily should be adopted. If we accept that all of the alternatives are potentially reasonable ways of thinking about characters and homology, then logic alone cannot help in determining which approach is most suitable. We can, however, compare them on the basis of heuristic success. If Nelson's approach is accepted, then all of the insights that flow from its application become potentially useful in cladistics. That these insights have led to a better understanding of relationships throughout the history of zoological systematics does not seem to be at issue (see Nelson 1973; Gould 1977). No doubt this approach will continue to provide new insights for systematists if it is not abandoned. The alternative approaches of Lundberg and de Queiroz, if accepted, simply destroy Nelson's argument and imply that the understanding gained from the study of ontogenetic character sequences may be irrelevant or misleading. Furthermore, they provide no alternative direct cladistic method for polarizing transformation series. In short, they suggest the replacement of a progressive line of research with one that promises only the lack of progress.

Alberch's criticism, mentioned above, is part of a more general examination of the developmental model underlying Nelson's argument. He characterizes this model as a causal sequence of discrete stages that is conserved during evolution. Alberch argues that the development of many, if not most features of real organisms is not consistent with this model. In particular he emphasizes a dynamic model of development that involves more complicated interactions between developmental parameters than the simple, causal sequence of stages in Nelson's model. That is, Nelson's developmental model is an oversimplification that is not applicable to the development of some (or most?) characters. This does not mean though that Nelson's method will produce spurious results but rather that it cannot be used in many cases. That is not a damning criticism of the method: subterminal evolutionary modifications of linear character sequences also result in ontogenies that are not fully informative. These "nonadditive" evolutionary changes would pose insurmountable problems to any systematic approach that relied exclusively on the direct method for the polarization of character transformation series. Neither Nelson nor any other advocate of his direct method has to my knowledge suggested such an approach. Alberch's criticism thus exposes the limitations, not any downfalls, of Nelson's method.

Probably the most common criticism leveled at the direct method is that homoplasy, especially paedomorphosis, may render its results partially or totally misleading. As Nelson (1978, 1985) himself has noted, however, homoplasy can mislead any systematic method. Rejecting the direct method for this reason is equivalent to rejecting cladistics on the basis that evolution may have proceeded "unparsimoniously," an argument that has been vigorously rebutted by numerous cladists.

CONCLUSIONS

I began by stating that my aim was to reexamine the role of parsimony in Nelson's (1973) methodology of cladistic analysis. I have argued that the analytical methods he proposed boil down to a system of three complementary, parsimony-based techniques. These are Nelson's direct and indirect methods for formulating tentative character phylogenies and comparative parsimony analysis for testing them. This system is a logically consistent approach to cladistic analysis given a suitable character concept. It is also self-contained to the extent that it does not rely on the results of other, mutually incompatible procedures such as phenetics, nor on assumptions concerning the adequacy of fossil samples.

These three analytical techniques may be aligned in a sequence representing their logical position in cladistic analysis. The direct method occupies the logically most fundamental position because only its application can provide synapomorphy hypotheses without recourse to a previous cladistic analysis. This does not mean that hypotheses derived using the direct method are infallible, only that they are the most basic synapomorphy hypotheses in cladistics. They are formulated within the narrowest frame of reference and rely minimally on accepted background knowledge. That aspect of comparative parsimony analysis concerned with testing hypotheses of synapomorphy occupies the next most basic position since it is applicable given a minimum of two different tentative character phylogenies for three taxa. The indirect method, outgroup comparison, occupies the least basic position because it operates within a broad framework, relying on corroborated background knowledge.

This logical sequence does not imply a corresponding historical sequence. In practice, cladistic analysis has not started from scratch but has worked within the framework of "traditional" classifications. These are a mixture of well corroborated clades along with residual paraphyletic taxa and some polyphyletic taxa (see, e.g., Eldredge and

Cracraft 1980:158–165). The history of cladistic systematics has largely involved the critical evaluation of traditional taxa and the reanalysis of those found to be unsupported by synapomorphies. This "cleaning up" procedure has involved a piecemeal application of all cladistic techniques. Some synapomorphies have been justified by the direct method and these have proved adequate in supporting large areas of cladistic knowledge. The congruence of characters with one another has given these areas enough internal cohesion to allow continued progress. However, other areas of cladistic knowledge such as the relationships of plants and unicellular organisms are not well endowed with direct support. It is here that the generalized direct method that I have proposed in this chapter could prove to be useful. It could provide an empirical basis that would prevent these parts of cladistic knowledge "toppling over" as unrooted networks.

ACKNOWLEDGEMENTS

I am grateful to Roger Carolin, Don Colless, Michael Crisp, Chris Humphries, Gareth Nelson, Colin Patterson, Norman Platnick, the late Donn Rosen, and Roberta Townsend for discussing with me their views on ontogeny and systematics. Humphries, Nelson, David Morrison, and Peter Stevens critically read an earlier draft of the manuscript. Louisa Murray prepared the figures.

REFERENCES

Alberch, P. 1985. Problems with the interpretation of developmental sequences. *Syst. Zool.* 34:46–58.

Brooks, D.R. and E.O. Wiley 1985. Theories and methods in different approaches to phylogenetic systematics. *Cladistics* 1:1–11.

Cambage, R. H. 1915–1928. *Acacia* seedlings I to XIII. *Proc. Roy. Soc. N.S.W.*, vols. 49,60 and 62.

Colless, D.H. 1967. The phylogenetic fallacy. *Syst. Zool.* 16:289–295.

Colless, D.H. 1969. The phylogenetic fallacy revisited. *Syst. Zool.* 18:115–126.

Cutter, E.G. 1969. *Plant Anatomy: Experiment and Interpretation. Part 1 Cells and Tissues.* London: Edward Arnold.

Eldredge, N. and J. Cracraft. 1980. *Phylogenetic Patterns and the Evolutionary Process.* New York: Columbia University Press.

Farris, J.S. 1971. The hypothesis of nonspecificity and taxonomic congruence. *Ann. Rev. Ecol. Syst.* 2:277–302.
Farris, J.S. 1982. Simplicity and informativeness in systematics and phylogeny. *Syst. Zool.* 31:413–444.
Farris, J.S. 1983. The logical basis of phylogenetic analysis. In N.I. Platnick and V.A. Funk, eds., *Advances in Cladistics,* 2:7–36. New York: Columbia University Press.
Farris, J.S., A.G. Kluge and M.J. Eckardt. 1970. A numerical approach to phylogenetic systematics. *Syst. Zool.* 19:172–189.
Felsenstein, J. 1982. Numerical methods for inferring evolutionary trees. *Quart. Rev. Biol.* 57:379–404.
Fletcher, J.J. 1920. On the correct interpretation of the so-called phyllodes of the Australian phyllodineous acacias. *Proc. Linn. Soc. N.S.W.* 45:24–47.
Gould, S.J. 1977. *Ontogeny and Phylogeny.* Cambridge: Harvard University Press, Belknap Press.
Hara, N. and D.R. Kaplan. 1980. The problem of phyllode dimorphism and homology in *Acacia verticillata. Bot. Soc. Amer. Ser. Publ.* 158:48.
Hennig, W. 1966. *Phylogenetic Systematics.* Urbana: University of Illinois Press.
Kaplan, D.R. 1975. Comparative developmental evaluation of the morphology of unifacial leaves in the monocotyledons. *Bot. Jahrb. Syst.* 95:1–105.
Kaplan, D.R. 1980. Heteroblastic leaf development in *Acacia:* Morphological and morphogenetic implications. *Cellule* 73:135–203.
Kaplan, D.R. 1984. The concept of homology and its central role in the elucidation of plant systematic relationships. In T. Duncan and T.F. Stuessy, eds. *Cladistics: Perspectives on the Reconstruction of Evolutionary History.* New York: Columbia University Press.
Kluge, A.G. 1985. Ontogeny and phylogenetic systematics. *Cladistics* 1:13–27.
Lipscomb, D.L. 1985. The eukaryotic kingdoms. *Cladistics* 1:127–140.
Løvtrup, S. 1978. On von Baerian and Haeckelian recapitulation. *Syst. Zool.* 27:348–352.
Lundberg, J.G. 1972. Wagner networks and ancestors. *Syst. Zool.* 21:398–413.
Lundberg, J.G. 1973. More on primitiveness, higher level phylogenies and ontogenetic transformations. *Syst. Zool.* 22:327–329.
Maddison, W.P., M.J. Donoghue and D.R. Maddison. 1984. Outgroup analysis and parsimony. *Syst. Zool.* 33:83–103.
Mayr, E. 1969. *Principles of Systematic Zoology.* New York: McGraw-Hill.
Mayr, E. 1982. *The Growth of Biological Thought: Diversity. Evolution, and Inheritance.* Cambridge University Press.
Nelson, G. 1973. The higher-level phylogeny of the vertebrates. *Syst. Zool.* 22:87–91.
Nelson, G. 1974. Historical biogeography: An alternative formalization. *Syst. Zool.* 23:555–558.
Nelson, G. 1978. Ontogeny, phylogeny, paleontology, and the biogenetic law. *Syst. Zool.* 27:324–345.

Nelson, G. 1985. Outgroups and ontogeny. *Cladistics* 1:29–45.

Nelson, G. and N. Platnick. 1981. *Systematics and Biogeography: Cladistics and Vicariance*. New York: Columbia University Press.

Patterson, C. 1981. Methods of paleobiogeography. In G. Nelson and D.E. Rosen, eds. *Vicariance Biogeography: A Critique*, pp. 446–489. New York: Columbia University Press.

Patterson, C. 1982. Morphological characters and homology. In K.A. Joysey and A.E. Friday, eds. *Problems of Phylogenetic Reconstruction*. pp. 21–74. London: Academic Press.

Philipson, W.R. 1965. The New Zealand genera of the Araliaceae. *N.Z. J. Bot.* 3:333–341.

Popper, K.R. 1961. *The Poverty of Historicism*. London: Routledge and Kegan Paul.

Popper, K.R. 1980. *The Logic of Scientific Discovery*. 10th (revised) impression. London: Hutchinson.

de Queiroz, K. 1985. The ontogenetic method for determining character polarity and its relevance to phylogenetic systematics. *Syst. Zool.* 34:280–299.

Rosen, D.E. 1982. Do current theories of evolution satisfy the basic requirements of explanation? *Syst. Zool.* 31:76–85.

Rosen, D.E. 1984. Hierarchies and history. In J.W. Pollard, ed., *Evolutionary Theory: Paths into the Future*, pp. 77–97. Chichester: Wiley.

Sachs, T. 1982. A morphogenetic basis for plant morphology. *Acta Biotheoretica*. 31:118–131.

Salmon, J.T. 1980. *The Native Trees of New Zealand*. Wellington: Reed.

Sattler, R. 1984. Homology—a continuing challenge. *Syst. Bot.* 9:382–394.

Seaman, F.C. and V.A. Funk. 1983. Cladistic analysis of complex natural products: developing transformation series from sesquiterpene lactone data. *Taxon* 32:1–27.

Sneath, P.H. and R.R. Sokal. 1973. *Numerical Taxonomy*. San Francisco: Freeman.

Stebbins, G.L. 1974. *Flowering Plants: Evolution Above the Species Level*. Cambridge: Harvard University Press.

Stevens, P.F. 1984. Metaphors and typology in the development of botanical systematics 1690–1960, or the art of putting new wine in old bottles. *Taxon* 33:169–211.

Tomlinson, P.B. 1982. Chance and design in the construction of plants. *Acta Biotheoretica* 31:162–183.

Vassal, J. 1979. Interêt de l'ontogénie foliaire pour la taxonomie et la phylogénie du genre *Acacia*. *Bull. Soc. Bot. Fr.* 126:55–65.

Voorzanger, B. and W.J. van der Steen. 1982. New perspectives on the Biogenetic Law? Syst. Zool. 31:202–205.

Wiley, E.O. 1981. *Phylogenetics: The Theory and Practice of Phylogenetic Systematics*. New York: Wiley.

3. The Characterization of Ontogeny

Arnold G. Kluge

Danser (1950:118) urged systematists to classify life cycles rather than stages in the ontogenetic continuum such as adults, juveniles, neonates, or embryos. His reason for doing so was simple—it is only natural to think of organisms in their entirety. Gould (1977) and Rosen (1982) amplified this argument by pointing out that ontogeny is more informative than any single life-history stage, because it is multidimensional. It includes direct and indirect information about biological time (Kluge and Strauss 1985). Illustrative of Danser's point, Patterson (1982:68) described the way the life cycle perspective solves the conundrum of "how a cranial capacity of 1200 cc can be characteristic of *Homo sapiens,* when no newborn individual has the character. The character of *H. sapiens* is not to have that cranial capacity, but both to lack it (general condition) and to have it (special condition), whereas all other organisms exhibit only the general condition."

While Danser's recommendation sounds appealing, because of the emphasis placed on having knowledge of the entire organism, it has been largely ignored (see, however, Nelson [1985], and elsewhere). Systematists continue to characterize the life cycle haphazardly, apparently in the absence of a formalism that simultaneously summarizes individual growth and differentiation. This chapter describes some recent advances in methodology that may lead to a synthesis of ontogenetic patterns (Kluge and Strauss 1985), followed by brief discussions of some of the more obvious philosophical and theoretical issues associated with implementing that formalism. My primary focus is the major advantages and disadvantages of treating species as life cycles.

BACKGROUND

The slowness with which Danser's suggestion has attracted attention may be due to the prominent role ontogenetic staging has played for so long in systematics (Gould 1977). For example, usually only adults are investigated when delimiting species, because they are likely to provide evidence of reproductive isolation. Subadults may exhibit differences, but they do not demonstrate evolutionary independence as convincingly as reproductively mature individuals. There is also the practical problem of what can be observed. Few ontogenetic transformations can be studied directly, and most inferences of continuity are generalizations derived from stages. Phylogenetic inference is another area of research where stages are recognized. Hennig, emphasizing their importance in his semaphoront concept, stated that the semaphoront "must be regarded as the element of systematics because, in a system in which the genetic relationships between different things that succeed one another in time are to be represented, we cannot work with elements that change with time. Accordingly the semaphoront corresponds to the individual in a certain, theoretically infinitely small, time span of its life, during which it can be considered unchangeable" (Hennig 1966:65). Over the years, Hennig's infinitely small time span qualification has been relaxed. Today, semaphoront is simply taken to mean an organism at a well-demarcated stage in its life–history, e.g., adult, juvenile, neonate, or embryo, during which time the characters of interest are invariant or insignificantly variable. Comparable semaphoronts are individuals at the same stage (Wiley 1981).

De Queiroz's (1985:292–97) opinion concerning the need to recognize semaphoronts was opposite that of Hennig's. Arguing from the point of view that phylogeny is a sequence of life cycles, de Queiroz concluded that the evidential basis for inferring historical relationships should be the ontogenetic transformations themselves, not the transformed elements (see OConnor [1984] for a case study). For example, the traditional representation of the alternative states of the parietal bone observed in adult tetrapods as paired or single, might be restated in terms of whether or not they fuse during ontogeny. Presumably, the rewording exercise is made nontrivial by being able to record additional rate-related variation. According to de Queiroz, application of the semaphoront concept constrains systematics to a study of "instantaneous" morphologies, which is the source of the confusion about paedomorphosis being exceptional to the ontogeny criterion. In fact, he went on to assert that "there can be no ontogenetic method for polarity determination; the ontogenetic transformations within a character tell nothing about the evolutionary polarity of that character relative to others (p. 292)."

De Queiroz's basic idea of ontogenetic transformations as characters is intriguing, even though many of his arguments are difficult to follow. In any case, I believe Danser's and de Queiroz's positions are complimentary, and would be identical if ontogeny was a uniserial unfolding of the phenotype. However, the life cycle is much more complex than that, and unlikely to be adequately summarized by ontogenetic transformations (*sensu* de Queiroz, 1985) analyzed piecemeal. For the sake of brevity and clarity I hereafter refer to characters as horizontal or vertical (Nelson 1985), instead of instantaneous morphologies (Hennig 1966) or ontogenetic transformations (de Queiroz 1985), respectively. The two types of characters are compared and contrasted in each of the discussions of philosophy and theory to follow, first in more traditional terms, then in the context of a model of growth and differentiation (Kluge and Strauss 1985).

The characterization of ontogeny will be considered only in the context of the principles of phylogenetic systematics (Hennig 1966). A major goal of this research program is the discovery of the species genealogy. Descent with modification (evolution) and the transmutation and reproductive isolation of species (speciation) are the major assumptions necessary to achieve that end. A historical entity, a common ancestral species and all lineages derived from it, is described by an evolved trait (apomorphy), a condition which its members have in common and which is not shared with closely related lineages (Farris and Kluge 1985a, b). Such a group of species is termed a clade, and a nested series of clades is referred to as a cladogram or genealogical hypothesis. A shared derived condition is a synapomorphy, a 1:1 correspondence or identity, and it serves as a preliminary proposition of homology. Plesiomorphy, the more generally distributed condition from which the apomorphy evolved, does not constitute evidence for common ancestry. The outgroup and ontogeny criteria provide the bases most often used to distinguish which condition observed among the ingroup (those taxa whose relationships are in question) is plesiomorphic. That variant observed in the outgroup, or that condition found more generally or earlier in ontogeny, is hypothesized to be primitive. Ultimately, the phylogeneticist chooses the most parsimonious cladogram as the best estimate of genealogy. It is in that context of maximum congruence among all synapomorphies that hypotheses of homology and homoplasy (convergence and evolutionary reversal) are finally deduced. The simplest phylogenetic hypothesis is most informative of the character distribution (Farris 1983), and its likelihood obtains from the probability it confers on the evolutionary transformation of traits (Sober 1985).

A MODEL OF GROWTH AND DIFFERENTIATION

All phenotypic variation appears to be continuous when examined at the level of populations, and nowhere is this more evident than during ontogeny (see below). Heterochrony and allometry are two of the best studied ontogenetic phenomena known to produce continuous variation, and there may be significant loss and distortion of that information when it is described qualitatively (Alberch, 1985). A general model is needed to represent all forms of growth and differentiation, throughout ontogeny, whatever their cause and in whatever scale they occur, continuous or discontinuous. The formalism recently developed by Alberch et al. (1979; see also Fink [1982], Gould [1977] and Kluge and Strauss [1985]), interrelating heterochrony and allometry, seems to meet the spirit of these requirements. Alberch, (1985) however, now interprets the model he helped to develop as not up to the task that I propose for it, as will be discussed below.

The Alberch et al. (1979) model, hereafter referred to as the model, consists of only two descriptive parameters, shape and size or age (figure 3.1). Both multivariate shape- and size-factors can be calculated easily and are used in the model because they more effectively distinguish those two aspects of form. A multivariate size-factor summarizes increase in several mensurable traits as an estimate of "general size." It is usually the major axis, first principal component (*PC1*), calculated from a co–variance matrix of log-transformed values. Defined in this manner, general size is that linear combination of morphometric traits that minimizes the mean square residual of all characters treated simultaneously. It provides the standard against which growth in different variables can then be compared, and such a plot is called an ontogenetic trajectory (figures 3.1–2).

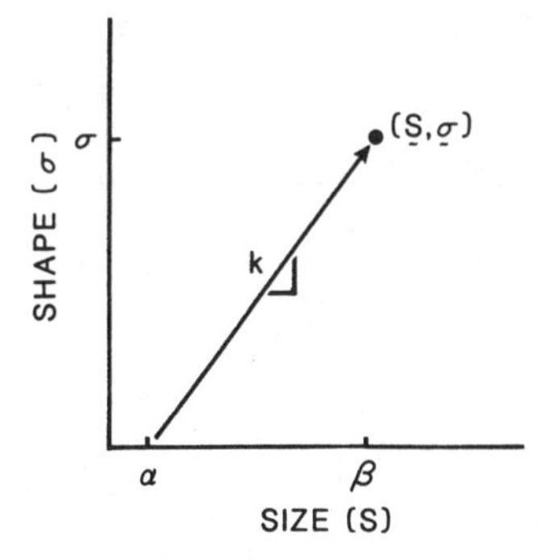

Figure 3.1. The growth trajectory of an organism as a function of the onset of growth (alpha), the cessation of growth (beta), and growth rate (k). The individual is described by shape and size-vectors at any point along the trajectory. Modified after Kluge and Strauss (1985).

Ideally, the trajectory summarizes the continuously changing physical appearance of the individual throughout its life. However, in most studies it will be a population sample or species composite, which then represents some average developmental pathway. Actual longitudinal studies of most organisms are impossible, because individuals can be measured only once due to destructive sampling. In these studies, the trajectories are only approximations of growth. Moreover, a trajectory will be limited

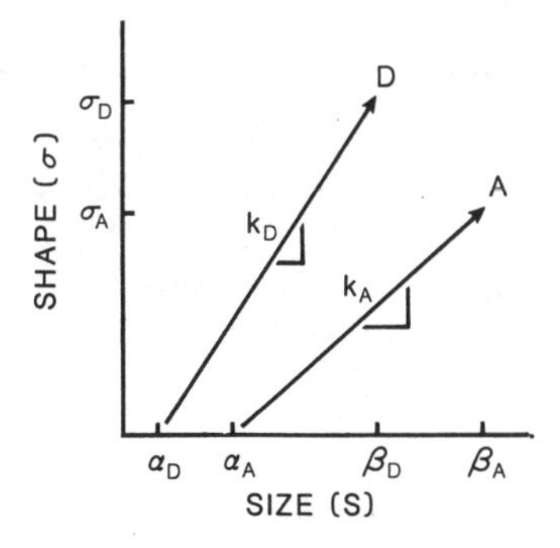

Figure 3.2. A comparison of two growth trajectories observed in an ancestral taxon (A) and its descendant (D). All three parameters, onset of growth (alpha), offset of growth (beta), and growth rate (k), have changed in the evolved form. Modified after Kluge and Strauss (1985).

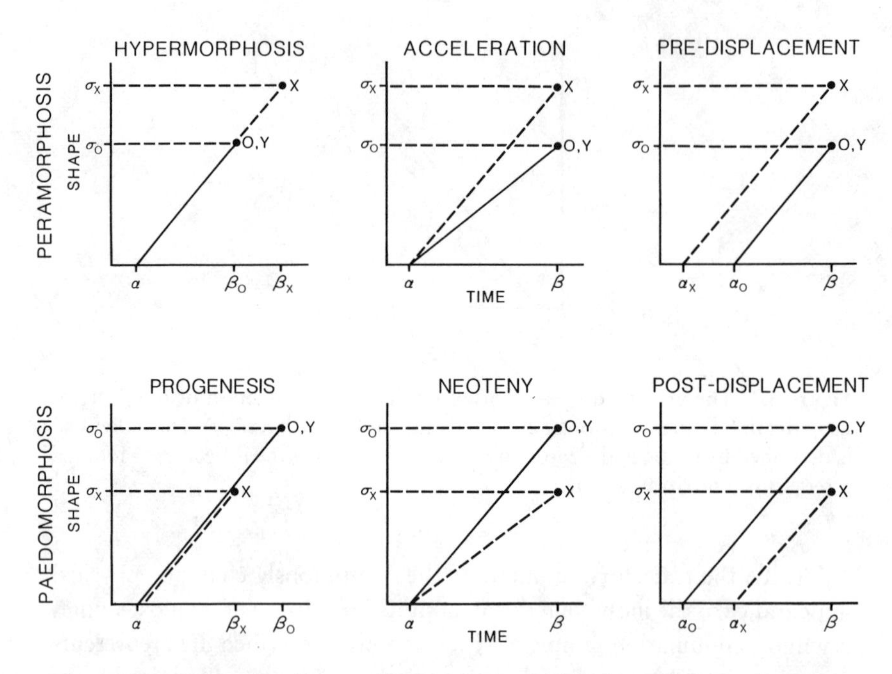

Figure 3.3. The six pure forms of heterochrony modeled in terms of variation in the onset of growth (alpha), the cessation of growth (beta), and growth rate (k). X and Y denote the trajectories of sister taxa whose relative primitiveness is to be decided by comparision with the trajectory of O, the outgroup.

to that portion of the life cycle during which comparable landmarks can be identified with a reasonable degree of confidence, as required by the analysis of size and shape (see, however, Strauss and Fuiman, 1985:1588). While it may be impossible to accurately describe a single trajectory from zygote to adult, the complete life cycle can be represented by several nonoverlapping trajectories that characterize all the different aspects of shape change. In vertebrates, most significant shape change begins with histogenesis, and there is probably little need to survey the early stages of ontogeny, through gastrulation.

Six pure cases of heterochrony, the change in relative timing of expression as a function of age, have been identified: progenesis, neoteny, and post-displacement are examples of paedomorphosis, and hypermorphosis, acceleration and pre-displacement are forms of peramorphosis (figure 3.3). Plotting ancestor and descendant trajectories together (*O*,

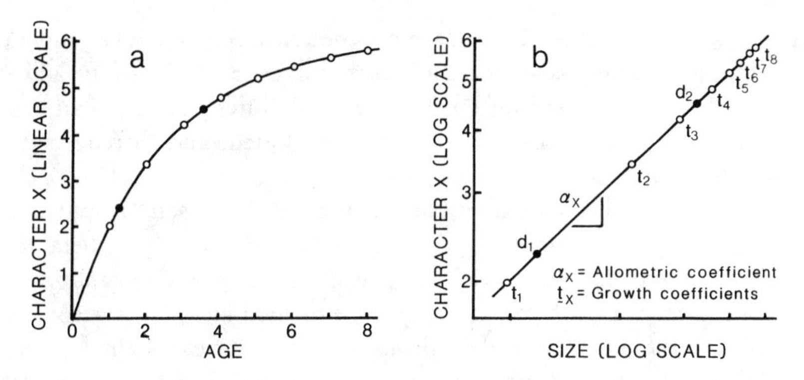

Figure 3.4. A mensural character x is plotted on the age-specific (a) and general size (b) determined growth trajectories. In a, open circles denote sizes at specific ages, closed circles timing of particular developmental events. Those sizes and events are represented on the trajectory in b. Modified after Kluge and Strauss (1985).

Y, and *X*, respectively) and separately varying the time of onset of growth (alpha), time of cessation of growth (beta), and growth rate *(k)* is sufficient to simulate those six cases. More realistic mixed cases of heterochrony can be produced by simultaneously varying two or all three of these developmental controls. An important feature of the model is that it accounts for terminal addition (peramorphosis or recapitulation) and deletion (paedomorphosis) without having to identify semaphoronts or discrete states. Hypotheses of heterochrony actually require at least a three taxon comparison, putative sister taxa, and one or more out groups, because ancestors and descendants cannot be identified and studied directly (figure 3.2; Fink 1982). The simultaneous comparison of ontogenetic trajectories in a historical context can be used to evaluate relative heterochrony, trait by trait, exposing the coupling and uncoupling of developmental events among species.

Once a trajectory has been plotted, both horizontal and vertical characters can be mapped on the curve (figure 3.4). Two discrete events are necessary to be able to accurately represent the position of a vertical character *(sensu* de Queiroz 1985), whereas only one is required to place the traditional horizontal character. One or a few nonoverlapping plots of shape change, on which many discrete size or age dependent changes are plotted, would provide a reasonably complete description of the gross features of a life cycle. A single ontogenetic trajectory might estimate growth, but numerous vertical and horizontal characters must be added to be able to approximate the full scope of differentiation. Different developmental processes and form can be plotted, and thereby better characterize the pattern of epigenesis.

The model yields several additional types of data useful to phylogeneticists, with variation in the developmental controls, alpha, beta and *k*, being the most obvious. While division of ontogeny into semaphoronts is unnecessary, growth trajectories can be divided into stages according to the onset and cessation of particularly obvious developmental events, and then compared in terms of the presence, absence, or relative positions of stages among taxa. All systematists may be more compelled to employ semaphoronts established in this manner, because a more precise study of ontogeny precedes their delimitation.

The cornerstone of the model is its analysis of allometry, the systematic change in shape with growth, and as such makes explicit variation in shape and shape change. A great deal of recent research has been devoted to improved methods for representing shape and shape change (Bookstein et al. 1985). However, the usefulness of these data in phylogenetic inference has not been explored extensively. The publications by Strauss and his associates (Bookstein et al. 1982; Strauss 1985; Strauss and Bookstein 1982; Strauss and Fuiman 1985) clearly illustrate the potential for further work in this area.

It is also possible to obtain additional information for phylogenetic inference by fitting quantitative models to the growth curves for individual traits, in the same way that the ontogenetic trajectories of multivariate vectors are employed. For example, Creighton and Strauss (1986) described the growth trajectories of a group of cricetine rodents in terms of the simple negative exponential model, which leads naturally to comparisons of linear trajectories (figure 3.5). The basic data in the cricetine study consisted of a set of such curves, one for each metric trait per taxon, from which alpha, beta and *k* were derived. Such univariate descriptions are consistent with the thesis that some, but not necessarily all, features of an organism are heterochronic. Additional information, such as age-specific growth rates and times at which particular developmental events occur, were mapped onto these curves. The end result was a matrix of growth and timing coefficients suitable for phylogenetic analysis. These few characters were then used to construct a cladogram, which was

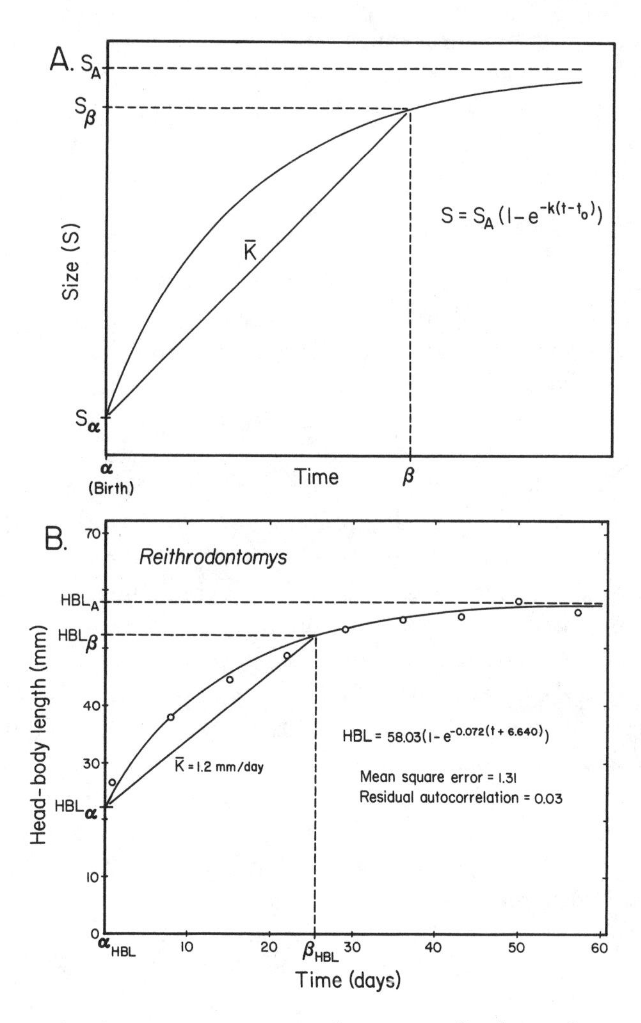

Figure 3.5. The descriptive parameters for a generalized negative exponential (Bertalanffy) growth curve (A), fitted to a curve of postnatal growth in the cricetine rodent *Reithrodontomys humulis* (B). The corresponding equation and statistics of fit for head and body length (HBL) are shown in B. From Creighton and Strauss (1986), with permission.

consistent with the major features of phylogenetic hypotheses based on more traditional types of data. Some novel insights into sistergroup relationships were discovered.

SURVEYING THE LIFE CYCLE

No stage in the life cycle seems to be immune to evolution (Kluge 1985), and systematists must therefore, broadly survey ontogeny for synapomorphies. Caenogenetic evolution will not be accounted for by examining only adults, and recapitulation will not be revealed by larvae alone. For example, Brooks and his collaborators (Brooks et al. 1985a, b; O'Grady 1985) found a high degree of congruence between cladograms of larval and adult parasitic flatworms and, not too surprisingly, synapomorphies from both stages gave the fullest resolution of cladistic relationships.

A more difficult survey problem relates to the multiserial patterns of ontogeny. Typically, many pathways of differentiation exist in each organism. In addition to the two or more primary routes, from zygote to adult, there are a large number of interconnecting secondary pathways (figure 3.6), and each exhibits one or more points of differentiation (figure 3.2; Kluge and Strauss 1985, fig. 9). Vertical characters should be more effective than horizontal characters at sampling novelties throughout ontogeny, because the former explicitly considers variation within as well as between pathways, whereas the latter does not. Moreover, vertical characters can be used to survey rate related variation, as can the model of growth and differentiation described above.

There is no reason to assume that evolutionary events will be strictly synchronized among pathways. Therefore, the number of semaphoronts that can be convincingly identified *a priori* will usually be too few to represent the variation present. In fact, the semaphoront concept will either limit the amount of variation that can be recorded (de Queiroz 1985), or its basic tenet of stage invariance will be violated. Empirical studies will be necessary to document the details of this tradeoff.

Not all evolution relevant to phylogenetic inference can be surveyed with vertical characters (Kluge 1985). For example, autopolyploidy is known to lead to speciation, with the derived lineages exhibiting few, if any, differences in embryology or adult form (Stebbins 1957). De Queiroz (1985) recognized this limitation to vertical characters and he recom-

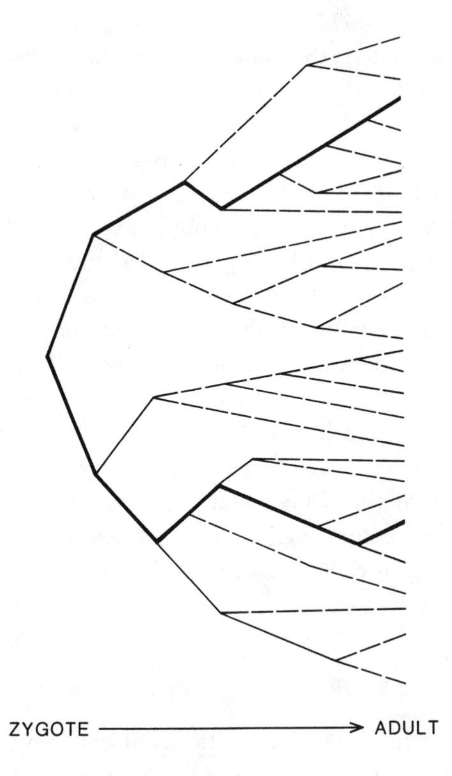

Figure 3.6. A highly simplified rendition of an individual developmental history, emphasizing the hierarchical system composed of organs, tissues, cells, enzymes, genes and DNA. When the pattern is purely bifurcating, the systematist is faced with having to sample two primary ontogenetic trajectories (solid lines) and a large number of minor ones (dashed lines).

mended employing horizontal characters when ontogenesis was absent or could not be observed.

These few remarks illustrate that neither vertical or horizontal characters have the potential to completely capture all systematically useful information. Since the growth and differentiation model (Kluge and Strauss 1985) combines the advantages associated with both it may be preferred. It provides a way of surveying allometric and heterochronic variation throughout the life cycle, and it is capable of representing novelties whose ontogenesis is not observed. Fitting quantitative growth curve models to the ontogenetic trajectories may even provide additional information. Unfortunately, the model is not without its own limitations. For example, the number of measurements that must be recorded in order to reflect differences in shape is inversely correlated with degree of similarity (R. Strauss, pers. comm.). Systematists studying distantly related species will see this as an advantage; those who work with closely related forms will have difficulty in distinguishing between the absence of interspecific shape differences and having recorded too few measurements. Ideally, the independent variable in the model should be ontogenetic time (figure 3.1). However, age is not always known, and size is regarded as a reasonably good approximation in most organisms. While there is much to recommend the use of a multivariate size-factor to estimate age (Bookstein et al. 1985), it is difficult to claim that it is a highly consistent measure among the taxa being compared without first assessing the contributions that each measurement makes to the general factor. Ideally, the within-group size vectors (e.g., PCI; Strauss and Fuiman 1985) should have positive loadings for all characters and account for a large amount of the total variance.

THE INDIVIDUALITY OF ONTOGENY

Interest in the relationship between patterns of ontogeny and phylogeny continues to increase (Bonner 1982; Goodwin et al. 1983; Raff and Kaufman 1983). However, few general mechanisms that might connect the two phenomena have been identified and most of those have not held up with further study. For example, Kauffman (1985) reviewed allometric, heterochronic, and quantitative genetic explanations for developmental constraint, and he concluded that none poses a constraining explanation. Why does a lasting understanding of such widespread and important phenomena as morphological stasis continue to evade us? Is the relationship between ontogeny and phylogeny so inherently difficult that our understanding is necessarily retarded, or are our units of comparison simply inappropriate (Alberch 1985)? Assuming that knowledge of history is important to understanding processes (Fink 1982), I believe it is at least worth considering the impediment to be the way systematists have traditionally characterized ontogeny.

Hull (1978) emphasized the importance of recognizing species as individuals, and he built a large part of his argument on analogies with organisms. Both a species and an organism are spatiotemporally localized, cohesive, and continuous entities, and as such perform special functions in nature. For example, the former is the unit of evolution, the latter has reproductive fitness and is the unit of selection. Hull concluded that to treat species as universals (classes) or to adopt Simpson's (1945) recommendation of dividing a single lineage which changes extensively through time into separate species is unnatural and destroys their special functions (Danser 1950). I believe cutting ontogeny into semaphoronts is equivalent to Simpson's dividing a lineage into chronospecies and I predict that it is just as likely to produce anomalies of observation and prediction. Perhaps, the homeostatic nature of an organism cannot be understood (Eldredge and Gould 1972) by arbitrarily dividing ontogeny into successive stages. If cuts are made only where there is rapid change, dramatic metamorphoses such as occur in arthropods, then arbitrariness may be minimized. However, when growth is even, as it is in many organisms, distinct stages are much less evident and the pattern of change is more accurately regarded as continuous. In any case, I doubt that cutting the ontogenetic continuum at any level can be considered totally nonarbitrary, regardless of the unevenness of growth, because heterochronic events are not necessarily synchronized across an ontogeny (Creighton and Strauss 1986; see above).

It is for these reasons that vertical characters might be favored over horizontal characters. The former emphasizes the natural continuity of ontogenesis, and therefore seems likely to contribute fewer anomalies to the study of developmental processes. Since the aforementioned model of growth and differentiation provides an even more coherent description of ontogenesis, it may be the preferred characterization. It can represent both continuous as well as discrete patterns of variation and it seems to provide the "dynamical framework" sought by Alberch (1985:56). Only long-term application of the model will reveal whether there exists improved understanding of process.

A CASE STUDY: PATTERNS OF CHANGE DURING ONTOGENY

Systematists are confronted with a variety of patterns of variation within and among populations, and a great deal of effort has been spent trying to discover the processes responsible for the continuities and discontinuities. While the genetic bases and epigenetic processes responsible for

	TERMINAL	NONTERMINAL
ADDITION	A – B – C / A – B – C – D*	A – B – C / A – B – D*– C
DELETION	A – B – C / A – B – *	A – B – C / A – * – C
SUBSTITUTION	A – B – C / A – B – D*	A – B – C / A – D*– C

Figure 3.7. Possible ontogenetic change (modified after O'Grady, 1985). A-D denote different phenotypic variants (states) which are representative of different semaphoronts in the life cycle; the transformation of one variant into another is implied by the connecting line; * designates the altered state in the transformation series; the primitive series is above the line, the derived series below it.

most transformations remain poorly delimited (see, however, Ambros and Horvitz [1984], and Oster and Alberch [1982]), it seems safe to assume that novel phenotypes of any magnitude are not produced *de novo*—the added state represents a modification, however small, of its predecessor (Raff and Kaufman, 1983; see, however, Alberch, 1985). Thus, when attempting to understand the origin of variation, the study of ontogeny and vertical characters may be critical. For example, heterochrony is thought to be an important cause of continuous change, and slight modifications in developmental rate are also believed to have a dramatic, macroevolutionary affect on the adult form. Allometry is believed to be another major cause of continuous variation.

According to O'Grady (1985; see also Brooks and Wiley, 1985), all phenotypic variation can be attributed to either addition, deletion or substitution of phenotypic states at some point in ontogeny (figure 3.7). Addition is illustrated as different conditions succeeding one another, deletion as loss of transformation. The basic notions of addition and deletion seem to apply just as well to continuous change; however, the examples are never so dramatic as those cited above. In such a presentation, horizontal characters are emphasized.

The numerous examples of terminal and nonterminal addition and deletion seem to give adequate testimony to the historical reality of these two types of change. There are many examples, among them milk producing glands in mammals (terminal addition), extraembryonic membranes, the amnion and allantois in tetrapods (nonterminal additions), limblessness in tetrapods, such as in gymnophionans (terminal deletions), and loss of the tadpole stage, as in direct developing frogs of the genus *Eleutherodactylus* (nonterminal deletion). The difference between substitution and addition is less evident. Substitution is illustrated as one phenotypic state replacing another during ontogeny, which is not unlike addition (figure 3.7; see also O'Grady, 1985). However, the apparent difference between substitution and addition can be identified only in a historical context (figure 3.8), and I interpret the difficulty I have had discovering unambiguous examples of substitution as evidence that it does not exist as a natural process. In fact, I doubt that substitution involves anything but a combination of additions and deletions. Even O'Grady's (1985) examples of substitution cannot be distinguished from addition and deletion. For example, he cited the evolution of the turtle appendicular skeleton as a case of nonterminal substitution, but it seems to be explained just as well as by terminal addition (Ruckes 1929; Walker 1947). Similarly, his scale-feather substitution example might just as well represent terminal addition (feathers). The apparent reality of substitution, in these and other cases, may be simply a consequence of presenting it in the context of a model of horizontal character variation. My point is that by using vertical characters, or the model of growth and differentiation, the reality of substitution might never have been entertained. Examining the origin of variation at the population level, viz., ontogenetically, might have revealed the true basis of the change.

ONTOGENY AND HOMOLOGY

"Only homology justifies the presumption of common ancestry and sistergroup relationships" (Kluge and Strauss 1985:257), and ontogenetic similarity is often used in formulating hypotheses of homology. According to Patterson (1982; see also Stevens 1984), phylogenetic systematists employ a taxic concept of homology; synapomorphies are judged to be homologs in terms of their ability to describe a historical entity. Taxic homology necessarily involves semaphoronts as the unit of comparison, because homologs are considered at particular levels of generality, phylogenetic and ontogenetic (Løvtrup 1978:351). There are no degrees of taxic homology; it is categorical, as a hypothesis.

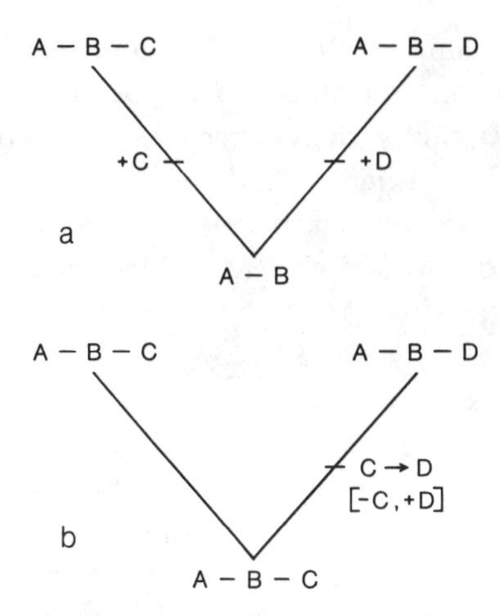

Figure 3.8. Two historical patterns of ontogenetic change. Independently acquired terminal additions are shown in a, whereas b describes causally related transformations, a deletion and addition, in the same ancestor-descendant lineage. Notice, both patterns involve addition and, according to fig. 3.7, only b qualifies as an example of substitution.

The conclusion that traits in different species are taxic homologs is reached after two sets of tests have been passed (Patterson 1982). The first of these, which has been referred to as pretesting, involves judgments as to conjunction and overall similarity—topographic, compositional, and ontogenetic. Conjunction occurs when putative homologs are observed in the same individual at the same time (a semaphoront), and this refutes the hypothesis of their homology (Hennig 1966). Logically, this test can be applied only when the comparison involves different states, such as crocodilian scales and avian feathers. It is meaningless to apply conjunction to feathers alone because there is no combination of states that would fail the test. Ontogeny is a time-ordered process, and similarity among pathways involves evaluations of temporal continuity, as well as the topographic and compositional aspects of the components in the transformation. Congruence means correlation among synapomorphies at a particular level of taxonomic generality on the most parsimonious genealogical hypothesis, and it provides the final test. Congruence is especially important, because the various types of nonhomology (homoplasy—convergence, parallelism, and evolutionary reversal) can pass the preliminary tests of conjunction and/or similarity but fail the ultimate test of congruence.

Rieppel (1985:23) claimed that "taxic homology means similarity based on topographic relations, a purely formal criterion, and does not entail material and temporal continuity." This definition misrepresents Patterson's concept of homology by stressing only topographic relations. It also ignores the fact that ontogenetic similarity involves transformation, and it also overlooks the material and temporal continuity of congruence—the most parsimonious hypothesis is adopted because of the probability it confers on the evolutionary transformation of traits (Sober 1985). Thus, Rieppel's dichotomy of being and becoming, pattern and process, is bridged by ontogeny and parsimony.

Ontogenetic similarity has long been known to be unnecessary in deducing homology. For example, Darwin (1872:466) stated that "community in embryonic structure reveals community of descent; but dissimilarity in embryonic development does not prove discommunity of descent, for in one of two groups the developmental stages may have been suppressed, or may have been so greatly modified through adaptation to new habits of life as to be no longer recognizable."

Numerous examples of taxic homology exist that by most accounts fail the test of ontogenetic similarity (Alberch 1985; Strauss and Fuiman 1985, figure 6). The blastocoels of chordates are considered homologs, even though the morphogenetic movements responsible for blastulation, infiltration or involution of the hypoblast, are markedly different (Manner 1964). Similarly, the neurocoels of all chordates are unquestionably homologous, but they too are formed by different processes, either cavitation or invagination (Nelson 1953).

Highly dissimilar morphogenetic movements are also involved in forming the chordate sclerocoel. Crocodilian scales and avian feathers are probably homologs at the level of Archosauria, but their development is markedly different (Maderson and Sawyer 1979). The cranial bones of gnathostomes provide several additional examples (de Beer 1937). For instance, the human parietal typically develops from four centers of ossification, rather than the usual two observed in other vertebrates. Thus, the taxic form of homology seems to employ ontogeny in a wise and conservative manner, given the prospect for nonterminal evolution. It is not unduly weighted over conjunction and other forms of similarity, and congruence is in place to serve as the ultimate arbiter.

Some comparative biologists view similarity of ontogeny as more important than all other criteria in deducing homology. A few even consider it both necessary and sufficient. I refer to this extreme position as ontogenetic homology. For example, Nelson (1978:335) concluded that "the mode of development itself is the most important criterion of homology," and more recently (1985:43), he claimed that homology is directly presented by ontogeny (see also de Queiroz 1985:294). In keeping with the emphasis he placed on developmental continuity, Nelson (1985) described the connection between homology and ontogeny as a vertical relation in order to distinguish it from the horizontal relation of taxic homology. Roth (1984:17) reached a similar conclusion concerning the importance of ontogeny, which is evident from her statement that "a necessary component of homology is *the sharing of a common developmental pathway.*"

A protocol for practicing the extreme form of ontogenetic homology is straightforward. For example, some bone in species *X* is homologous with some bone in species *Y* if they exhibit the same patterns of skeletogenesis. The bones are not homologous if the pathways are dissimilar. Since ontogenetic similarity is presented as an infallible parameter with which to deduce homologs there is no call for testing it with conjunction, topographic, and compositional similarity, or congruence. Such a position is like the transcendental morphologists and their claim of essential similarity (Geoffroy Saint Hilaire 1818; Owen 1848), and equally vulnerable to criticism. Most importantly, the connection between the ontogenetic and phylogenetic levels of generality is imperfect. The number of examples of nonterminal additions and deletions that have been described is simply too large for all of them to be attributed to misinterpretation. Therefore, the extreme form of ontogenetic homology is of little interest to practicing systematists simply because it is unrealistic. Some form of testing is necessary, and ultimately, like taxic homology, that involves congruence.

Much of the literature in which ontogenetic homology is said to be practiced is difficult to interpret. The problem involves a failure to distinguish between ontogeny as an *explanation* of homology and ontogeny as a *criterion* involved in its recognition (Kluge and Strauss 1985; Mason 1957; Stevens 1984). I believe this failure is especially significant for phylogenetic systematics because it confuses *process* with *pattern*. Developmental pathway, mode of development, and vertical relation are considerations of process, while the notion of sharing is an essential part of most criteria used to deduce cladistic patterns (Nelson 1985; Roth 1984). Process and pattern are different, and mixing the two seems unlikely to allow a clear and independent assessment of either phenomenon. This is particularly critical when it comes to the study of processes because our "understanding" may be markedly affected by the phylogenetic pattern we have in mind. For example, Fink (1982) demonstrated that distinguishing among the six pure cases of heterochrony (figure 3.3) requires a historical hypothesis. In mixing process and pattern, there is also the potential for circular reasoning and making untestable claims.

Adopting de Querioz's (1985) vertical character concept does not require a significant adjustment in our thinking about taxic homology. The fact that the ontogenetic transformation is the character does not involve a tautology anymore than does using topographic and compositional similarity in pretesting render the horizontal character concept tautologous. Moreover, the vertical, like the horizontal, character concept allows for assessments of conjunction and topographic and compositional similarity. The test of conjunction is failed if putative homologous pathways are found in the same ontogeny. Degree of topographic similarity is judged in terms of the relation among different pathways in a single ontogenetic hierarchy (figure 3.2), and compositional similarity requires assessing the anatomical likeness of the more discrete components in the ontogenetic transformations being compared.

The concept of taxic homology also applies without difficulty to the model of growth and differentiation described earlier (figure 3.4; see, however, Bookstein et al. 1985). The vertical or horizontal characters that are mapped on to the ontogenetic trajectory present no problem; they are tested in the manner described by Patterson (1982). Likewise, the landmarks used in the measurement of the size and shape axes are amenable to Patterson's two levels of testing. In fact, errors in estimation of size and shape may be minimized by employing multivariate factors. A potential limitation to the model concerns those cases where congruence indicates putative homologues are homoplasies. While homoplasy may be the simplest explanation, the systematist may also wish to rule out the possibility of clerical or measurement error. This may be difficult or impossible because the original observations have been transformed or are several analytical steps removed from the character states actually used in formulating the cladistic hypothesis. For example, suppose a discrete character mapped on a trajectory is interpreted as independently evolved. How is one to determine the possibility of error of estimation in the multivariate size and shape-factors?

POLARIZATION AND THE ONTOGENY CRITERION

The units of classification in phylogenetic systematics, be they species or more inclusive groups, are delimited in terms of character evidence, vertical or horizontal. Logically, all such characterizations are relative; taxa either exhibit a condition or they do not. Since only synapomorphies constitute evidence of sistergroup relationships some basis must be employed for ordering the comparable (p or not p) states in an ancestor-descendant series. The ontogeny criterion, like the outgroup criterion, provides a reasonable basis on which those states can be ordered (see Kluge 1985, and Kluge and Strauss 1985, for limitations). The former criterion has been formulated in terms of strict ontogenetic precedence (Hennig 1966); the earlier appearing state is primitive, or in terms of generality among ontogenies, the more widespread condition is primitive (Nelson, 1978).

De Querioz (1985:280), in promoting the vertical character concept, concluded that "there can be no ontogenetic method." His argument was that the ontogenetic transformation is the character, and that the ontogeny criterion therefore pertains to precedence within pathways (Hennig 1966), not generality among pathways (Nelson 1978). If generality among pathways is accepted, then the ontogeny criterion is available for polarizing vertical character states. The latter is problematical because it is equivalent to assuming common equals primitive (Farris and Kluge 1979; Kluge 1985). Thus, the outgroup criterion is recommended in the analysis of vertical characters. Similarly, as discussed in the context of heterochrony (figure 3.3), the outgroup criterion is generally the preferred method of polarization.

DISCUSSION

I am at a loss to explain Alberch's (1985) criticisms of his own model of growth and differentiation (Alberch et al. 1979), which I promote here. His criticisms don't seem to be relevant. For example, he now imagines that the model requires the recognition of discrete states and that it can't accommodate the dynamic framework of development that he believes exists. His first point is not true. While the model has been presented in terms of its ability to simulate the six pure cases of heterochrony (figure 3.3), it can represent much more complex and subtle forms of rate related variation on a continuous scale (Kluge and Strauss 1985). Alberch's second claim is also untrue, to the extent that there are observable *patterns* of form and process. I have no reason to doubt his conclusion that ontogenetic sequences are not necessarily conserved (see above), and that processes are. However, a more precise and convincing picture should become evident when the patterns of both developmental processes and form are mapped on the same ontogenetic trajectory. Until such comparisons are actually made I am forced to discount Alberch's claims of special knowledge of developmental processes, just as I do those made by evolutionary systematists concerning natural selection, adaptation and adaptive zones, convergence and parallelism, and the like.

I believe there are several advantages to describing the life cycle in terms of the model of growth and differentiation set forth by Kluge and Strauss (1985). De Queiroz's vertical character concept is relatively incomplete and it doesn't appear to offer any unique advantage over the model. Perhaps most important, the model is capable of representing all forms of variation, terminal and nonterminal, continuous and discrete. Its quantitative nature leads to greater precision, and the arbitrary division of the ontogenetic continuum into semaphoronts is avoided altogether, or entertained after the life cycle has been described. The model incorporates information about the timing of differentiation events, as well as specific "growth laws." The study of mixed cases of heterochrony is easily accomplished with the model, a more difficult or impossible task when using discrete characters in the context of semaphoront comparisons. Change in shape can be deduced from relative allometries. It simultaneously characterizes relative and absolute growth rates of individual morphological traits (change with age), and covariation among suites of traits, both continuous and discrete (change with size). The presence of discontinuities in the growth trajectory draws attention to subtle caenogenetic changes.

The special insights into pattern and process that may come from employing the model of growth and differentiation can be discovered only with more empirical research. Unfortunately, these studies may be a long time in coming. I suspect most systematists will find excuses in the extra burden of quantification and statistics that are necessary to estimate size and shape. Others will see themselves as having no opportunity to characterize ontogenetic continua, because most collections are instantaneous samples and the characters employed require destructive sampling. However, let me emphasize that preliminary studies (e.g., Creighton and Strauss 1986) suggest that the problems are not as difficult as they may seem and that novel insights are waiting to be made with the new characterization of ontogeny.

ACKNOWLEDGMENTS

Most of my understanding of ontogeny came from the Tuesday evening meetings of the Systematics Discussion Group at the University of Michigan. I especially wish to thank Jacques Gauthier, Fred Kraus, and Rich Strauss for giving freely of their time and ideas. The National Science Foundation, grant BSR 83–04581, provided financial assistance.

REFERENCES

Alberch, P. 1985. Problems with the interpretation of developmental sequences. *Syst. Zool.* 34:46–58.
Alberch, P., S. J. Gould, G. F. Oster, and D. B. Wake. 1979. Size and shape in ontogeny and phylogeny. *Paleobiology* 5:296–317.
Ambros, V., and H. R. Horvitz. 1984. Heterochronic mutants of the nematode *Caenorhabditis elegans. Science* 226:409–416.
Bonner, J. T. 1982. *Evolution and Development.* New York: Springer-Verlag.
Bookstein, F. L., B. Chernoff, R. L. Elder, J. M. Humphries, G. R. Smith, and R. E. Strauss. 1985. Morphometrics in evolutionary biology: The geometry of size and shape change, with examples from fishes. *Spec. Publ. Philad. Acad. Nat. Sci.* 15:1–277.
Bookstein, F. L., R. E. Strauss, J. M. Humphries, B. Chernoff, R. L. Elder, and G. R. Smith. 1982. A comment upon the uses of Fourier methods in systematics. *Syst. Zool.* 31:85–92.
Brooks, D. R., R. T. O'Grady and D. R. Glen. 1985a. The phylogeny of the Cercomeria Brooks, 1982 (Platyhelminthes). *Proc. Helminthol. Soc. Wash.* 52: 1–20.
Brooks, D. R., R. T. O'Grady and D. R. Glen. 1985b. Phylogenetic analysis of

the Digenea (Platyhelminthes: Cercomeria) with comments on their adaptive radiation. *Can. J. Zool* 63:411–443.

Brooks, D. R. and E. O. Wiley. 1985. Theories and methods in different approaches to phylogenetic systematics. *Cladistics* 1:1–11.

Creighton, G. K. and R. E. Strauss. 1986. Comparative patterns of growth and development in cricetine rodents and the evolution of ontogeny. *Evolution* 40:94–106.

Danser, B. H. 1950. A theory of systematics. *Bibliotheca Biotheoretica* 4:117–180.

Darwin, C. R. 1872. *On The Origin of Species, with Additions and Corrections.* London: Murray.

De Beer, G. R. 1937. *The Development of the Vertebrate Skull.* Oxford: Clarendon Press.

De Queiroz, K. 1985. The ontogenetic method for determining character polarity and its relevance to phylogenetic systematics. *Syst. Zool.* 34:280–299.

Eldredge, N. and S. J. Gould. 1972. Punctuated equilibria: An alternative to phyletic gradualism. In T. J. M. Schopf, ed., *Models in Paleobiology,* pp. 82–115. San Francisco: Freeman, Cooper.

Farris, J. S. 1983. The logical basis of phylogenetic inference. In N. I. Platnick and V. A. Funk, eds., *Advances in cladistics, volume 2: Proceedings of the Second Meeting of the Willi Hennig Society,* pp. 7–36. New York: Columbia University Press.

Farris, J. S. and A. G. Kluge. 1979. A botanical clique. *Syst. Zool.* 28:400–411.

Farris, J. S. and A. G. Kluge. 1985a. Parsimony, synapomorphy, and explanatory power: A reply to Duncan. *Taxon* 34:130–135.

Farris, J. S. and A. G. Kluge. 1985b. Synapomorphy, parsimony and evidence. *Taxon* 35:298–306.

Fink, W. L. 1982. The conceptual relationship between ontogeny and phylogeny. *Paleobiology* 8:254–64.

Geoffroy Saint-Hilaire, E. 1818–22. *Philosophie anatomique.* Paris: J. B. Bailliere.

Goodwin, B. C., H. Holder and C. C. Wylie. 1983. *Development and Evolution.* Cambridge: Cambridge Univ. Press.

Gould, S. J. 1977. *Ontogeny and Phylogeny.* Cambridge: Harvard Univ. Press.

Hennig, W. 1966. *Phylogenetic Systematics.* Urbana: Univ. Illinois Press.

Hull, D. L. 1978. A matter of individuality. *Philos. Sci.* 45:335–360.

Kauffman, S. A. 1985. New questions in genetics and evolution. *Cladistics* 1:247–265.

Kluge, A. G. 1985. Ontogeny and Phylogenetic Systematics. *Cladistics* 1:13–27.

Kluge, A. G. and R. E. Strauss. 1985. Ontogeny and systematics. *Ann. Rev. Ecol. Syst.* 16:247–268.

Løvtrup, S. 1978. On von Baerian and Haeckelian recapitulation. *Syst. Zool.* 27:348–352.

Maderson, P. F. A. and R. H. Sawyer. 1979. Scale embryogenesis in birds and reptiles. *Anat. Rec.* 193:609.

Manner, H. W. 1964. *Elements of Comparative Vertebrate Embryology*. New York: Macmillan Co.
Mason, H. L. 1957. The concept of the flower and the theory of homology. *Madrono* 14:81–95.
Nelson, G. 1978. Ontogeny, phylogeny, paleontology, and the biogenetic law. *Syst. Zool.* 27:324–345.
Nelson, G. 1985. Outgroups and ontogeny. *Cladistics* 1:29–45.
Nelson, O. E. 1953. *Comparative embryology of the vertebrates*. New York: Blakiston Co.
OConnor, B. M. 1984. Phylogenetic relationships among higher taxa in the Acariformes, with particular reference to the Astigmata. In D. A. Griffiths and C. E. Bowman, eds. *Acarology VI,* 1:19–27. Chichester: Ellis Horwood.
O'Grady, R. T. 1985. Ontogenetic sequences and the phylogenetics of parasitic flatworm life cycles. *Cladistics* 1:159–170.
Oster, G. and P. Alberch. 1982. Evolution and bifurcation of developmental programs. *Evolution* 36:444–59.
Owen, R. 1848. *On the Archetype and Homologies of the Vertebrate Skeleton*. London: R. and J. E. Taylor.
Patterson, C. 1982. Morphological characters and homology. In K. A. Joysey and E. A. Fridays, eds. *Problems of Phylogenetic Reconstruction,* pp. 21–74. London: Academic Press.
Raff, R. A. and T. C. Kaufman. 1983. *Embryos, Genes, and Evolution*. New York: Macmillan.
Rieppel, O. 1985. Ontogeny and the hierarchy of types. *Cladistics* 1:234–246.
Rosen, D. E. 1982. Do current theories of evolution satisfy the basic requirements of explanation? *Syst. Zool.* 31:76–85.
Roth, V. L. 1984. On homology. *Biol. J. Linn. Soc.* 22:13–29.
Ruckes, H. 1929. Studies of chelonian osteology. Part II. The morphological relationship between the girdles, ribs and carapace. *Ann. New York Acad. Sci.* 31:81–120.
Simpson, G. G. 1945. The principles of classification and a classification of mammals. *Bull. Am. Mus. Nat. Hist.* 85:1–350.
Sober, E. 1985. A likelihood justification of parsimony. *Cladistics* 1:209–233.
Stebbins, G. L., Jr. 1957. *Variation and Evolution in Plants*. New York: Columbia University Press.
Stevens, P. F. 1984. Homology and phylogeny: Morphology and systematics. *Syst. Bot.* 9:395–409.
Strauss, R. E. 1985. Static allometry and variation in body form in the South American catfish genus *Coydoras* (Callichthyidae). *Syst. Zool.,* 34:381–396.
Strauss, R. E. and F. L. Bookstein. 1982. The truss: Body form reconstruction in morphometrics. *Syst. Zool.* 31:113–135.
Strauss, R. E. and L. A. Fuiman. 1985. Quantitative comparisons of body form and allometry in larval and adult Pacific sculpins (Teleostei: Cottidae). *Can. J. Zool.* 63:1582–1589.

Walker, W. F., Jr. 1947. The development of the shoulder region of the turtle, *Chrysemys picta marginata,* with special reference to the primary musculature. *J. Morph.* 80:195–250.

Wiley, E. O. 1981. *Phylogenetics: The Theory and Practice of Phylogenetic Systematics.* New York: Wiley-Interscience.

4. The Systematic Implications of Pollen and Spore Ontogeny

Stephen Blackmore and Peter R. Crane

The resistant sporopollenin wall of pollen grains and spores has been a rich source of characters in systematic botany for over 150 years (Wodehouse 1935; Erdtman 1952, 1969; Blackmore 1984). Attempts to explain this diversity in phylogenetic terms have employed a variety of approaches, but frequently "trends" in pollen and spore evolution have been proposed with little discussion of the principles from which they have been developed. With the application of cladistic techniques to plant systematics (Koponen 1968; Bremer and Wanntorp 1978; Funk and Stuessy 1978; Humphries 1979; Hill and Crane 1982; Humphries and Funk 1984) more explicit approaches to determining character phylogenies have been adopted and palynological data have been increasingly integrated into more inclusive character analyses (Hill and Crane 1982; Bolick 1983; Crane 1985a,b). In cladistics the most widely used approach to resolving the direction of change in character transformations is outgroup comparison (Watrous and Wheeler 1981; Stevens 1980; Wiley 1981) whereby: "Given two characters that are homologues and found within a single monophyletic group, the character that is also found in the sister group is the plesiomorphic character whereas the character found only within the monophyletic group is the apomorphic character" (Wiley 1981:139). In palynology outgroup comparison has been applied either implicitly (Walker 1976, Walker and Doyle 1975) or explicitly (Blackmore 1982; Blackmore and Cannon 1983; Kress and Stone 1983; Linder and Ferguson 1985; Kress 1986) in studies at a variety of taxonomic levels. In some cases the results have been corroborated by comparison with the extensive fossil record of pollen and spores (Muller 1970; Doyle and Hickey 1976), but with few exceptions (Abadie and Hideux 1979; Hideux and Abadie 1981, 1985, 1986; Stone et al. 1981; Kress and Stone 1982; Blackmore 1984) such comparisons have rarely been extended to include ontogenetic information. Indeed the relevance of ontogenetic data to elucidating patterns of pollen evolution has scarcely been considered.

Since the advent of electron microscopy, knowledge of pollen and spore development has dramatically increased and a range of different taxa, including fossils (Taylor and Rothwell 1982; Taylor and Alvin 1984), have been examined (Heslop-Harrison 1968; Dickinson 1976; Neidhart 1979; Lugardon 1976; Audran 1981). Although synthesis of the current scattered literature is hampered by terminological complexities and by the variety of investigative techniques employed, in our view a general ontogenetic explanation of the major features of pollen and spore morphology is beginning to emerge. Furthermore, recently devised techniques for studying pollen ontogeny with the scanning electron microscope (Barnes and Blackmore 1984a,b, 1986; Blackmore and Barnes 1985) greatly facilitate comparative ontogenetic studies.

We consider here the application of ontogenetic data to systematic and phylogenetic problems in palynology. First we describe the major stages in pollen and spore ontogeny, and establish a framework within which to compare developmental processes in different taxa. We then review developmental data relating to five suites of palynological characters; dispersal unit, ectexine and endexine, reduced exines, apertures, and caveate and saccate pollen. We outline the ontogenetic basis for the major, systematically important, variations in these features. Finally, we discuss the general applicability of ontogenetic studies to clarifying phylogenetic patterns in pollen and spore diversity, particularly the utility of developmental data in generating hypotheses of character polarity.

MAJOR STAGES IN POLLEN AND SPORE ONTOGENY

The development of spores and pollen (the microspores of seed plants; Chaloner 1970) takes place during a critical phase in the life-cycle of land plants spanning the transition between the diploid sporophyte and the haploid gametophyte generation. In addition to the major genomic chromosomal and cytoplasmic changes associated with meiosis the production of pollen or spores from their mother cells involves the generation of a complex cell wall, the sporoderm, that constitutes the interface between the gametophyte and its environment during dispersal. Usually the outer part of the spore or pollen wall, the exine, is composed largely of sporopollenin—an extremely resistant class of carotenoid and/or carotenoid ester polymers (Shaw 1971; Southworth 1973). The inner pecto-cellulosic intine is more like a typical plant cell wall. It is the exine that has provided most of the systematically important palynological data and in this paper we are concerned almost exclusively with the ontogenetic processes leading to exine deposition.

Pollen and spore development takes place within a sporangium and cell wall substances are synthesised by specialized sporangial tissues as well as internally by the developing pollen grain or spore. Thus, although there are some superficial similarities to the development of unicellular organisms, pollen and spore wall deposition is controlled by the interaction of both internal and external processes. The complete ontogeny of the pollen or spore wall is a continuous process but can be divided for convenience into three main stages. In figure 4.1 the development of angiosperm and gymnosperm pollen grains is shown in diagrammatic form with the free spore stage subdivided to show additional detail.

Meiosis

The meiotic division marks the beginning of the gametophyte generation. Prior to division the mother cells begin to secrete a "special cell wall" that varies in composition in different groups of plants (Pettitt 1971; Neidhart 1979). It is composed of callose, a β–1,3–linked glucan found in spermatophytes, certain bryophytes, and some pteridophytes. In other bryophytes and pteridophytes (*Lycopodium* and eusporangiate ferns) the special cell wall possesses different staining properties (Pettitt 1971; Pacini et al. 1985). Technical difficulties involved with the detection and staining of callose (Smith and McCully 1978) make the significance of these differences difficult to assess. The special cell wall isolates the mother cells from the influence of the surrounding tissues of the sporangium. Synthesis of the special cell wall continues throughout meiosis, uniting the meiotic products in a tetrad and also generally separating them from each other. The geometry of the tetrad is influenced by the meiotic spindle and the timing of the two meiotic divisions and is closely

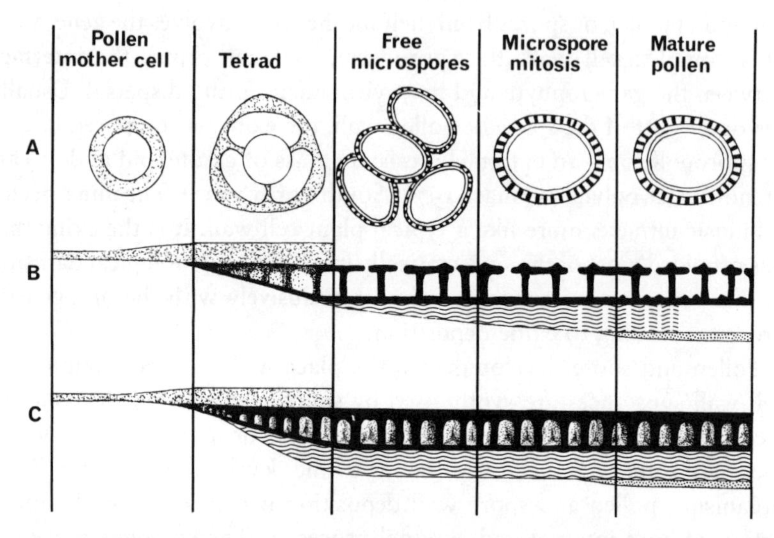

Figure 4.1. Principal stages in pollen development. Diagrams of walls in outline (A) and in sections for angiosperms (B) and gymnosperms (C). The pollen mother cell secretes a callose wall (stippled) present until the end of the tetrad stage. While in tetrads a primexine (black in B and C) is deposited and begins to differentiate into ectexine, endexine (wavy lines) deposition begins and may be completed. Callose wall digestion marks the start of the free microspore stage, during which ectexine and endexine deposition may continue. By microspore mitosis the exine is complete; between this stage and maturity the intine (dotted) is deposited and the endexine of angiosperms loses its lamellated appearance.

correlated with the number and position of germinal apertures (Huynh 1973, 1976a,b). The mechanism by which this influence is exerted remains unclear but it may involve gradients of cytoplasmic substances (Ford 1971), physical spacing effects (Wodehouse 1935; Kuprianova 1979; Melville 1981) or be directly related to the orientation of spindle microtubules. In addition to isolating the meiotic products the special cell wall may also function in maintaining tetrad geometry and controlling the generation of exine pattern (Barnes and Blackmore 1986).

The tetrad stage

This stage extends from the second division of meiosis to the enzymatic dissolution of the special cell wall, at which point the microspores usually disassociate. The deposition of the first components of the sporoderm is initiated during the tetrad stage. Polysaccharide and sporopollenin precursors are deposited between the special cell wall and the plasma membrane. This material, referred to as primexine (Heslop-Harrison 1963), is considered to form a template upon (or within) which sporopollenin accretion and polymerisation occurs (Heslop-Harrison 1963, 1968; Rowley and Skvarla 1975). In certain plants deposition of sporopollenin precursors and their polymerization may be virtually simultaneous so that a primexine is not always considered to be present (e.g. *Zamia*, Zavada 1983a). Since enucleate cytoplasmic fragments can form normal walls (Rogers and Harris 1969; Knox 1984) the production and organization of the primexine must be controlled, at least in part, by the spore cytoplasm and not the nucleus. At a later stage, usually shortly before the dissolution of the special cell wall, there is further sporopollenin deposition on characteristic white-line centered lamellae (Rowley and Southworth 1967) of unit membrane dimensions between the plasma membrane and the outer wall layers. This inner component of the exine is generally less electron dense in transmission electron microscopy and in most groups is identified as the endexine.

The free spore stage

This stage begins with the enzymatic dissolution of the special cell wall. Once liberated from the callose the developing spores or pollen grains may undergo marked expansion, often resulting in changes in shape and ornamentation. Substances in the sporangial locule may also be incorporated directly into the wall. Tapetal cells may become highly active and secretory, either maintaining their cellular integrity or losing their cell walls and forming a periplasmodium (Maheshwari 1950; Pacini et al. 1985). Subsequently, during the final maturation of the pollen grains or spores, the intine is deposited and there are important changes in the cytoplasm. Plastids or starch grains may accumulate, and the cytoplasm usually becomes dehydrated, a condition that marks the onset of a period of dormancy (Heslop-Harrison 1979), and that aids survival during dispersal (Blackmore and Barnes 1986). Bryophyte and pteridophyte spores, in contrast, may become vacuolate and photosynthetic at maturity. The breakdown of sporangial cells may produce a variety of surface coatings (Dickinson and Lewis 1973; Pettitt and Jermy 1974). In angiosperms this "pollenkitt" may include recognition substances (Heslop-Harrison 1976) and form an adhesive that binds pollen grains together during dispersal (Pankow 1957; Hesse 1978). In developing spores mitotic division of the microspore nucleus is often delayed until germination, but in pollen grains this division usually precedes development of the intine. "Microspore mitosis" and "post-mitotic maturation" are therefore often considered (as in fig. 4.1) as two additional stages in pollen development (eg. Blackmore and Barnes 1985).

The major features of pollen and spore development outlined above, although highly simplified, are common to the ontogeny of almost all pollen and spores that have been investigated (Knox 1984). It is variation in the details of these processes that is responsible for the considerable morphological diversity of pollen and spores. Five suites of characters dispersal unit, ectexine and endexine, reduced exines, apertures and saccate or caveate pollen have been selected for detailed discussion to exemplify the outcome of variations in ontogeny.

DISPERSAL UNIT

The majority of pollen grains and spores are dispersed individually, as monads. However, the occurrencc of permanent tetrads, corresponding to the products of a single pollen or spore mother cell, is widespread through a range of monocotyledonous and dicotyledonous families (see Erdtman 1945, and Walker and Doyle 1975 for reviews) and in certain pteridophytes (e.g., *Selaginella,* Erdtman and Sorsa 1971) and Bryidae (e.g., *Andreaea,* Erdtman 1965). With even a simple understanding of relationships in flowering plants it is clear that the character has arisen many times. It is an excellent example of convergence and has been shown to be an effective reproductive strategy accomplishing simultaneous fertilization of many ovules by the transfer of a single pollen unit in taxa with numerous ovules (Cruden 1977; Kenrick and Knox 1982; Knox and Kenrick 1983; Knox 1984). Permanent tetrads arise as a result of varying degrees of fusion of the pollen or spore walls (Skvarla and Larson 1963, Knox 1984) that are readily explained as consequences of relatively minor ontogenetic changes.

Usually pollen or spore mother cells secrete callose, which invests the tetrad in a common special cell wall (Heslop-Harrison 1966; Longly and Waterkeyn 1979; Hideux and Abadie 1981). As the four daughter cells

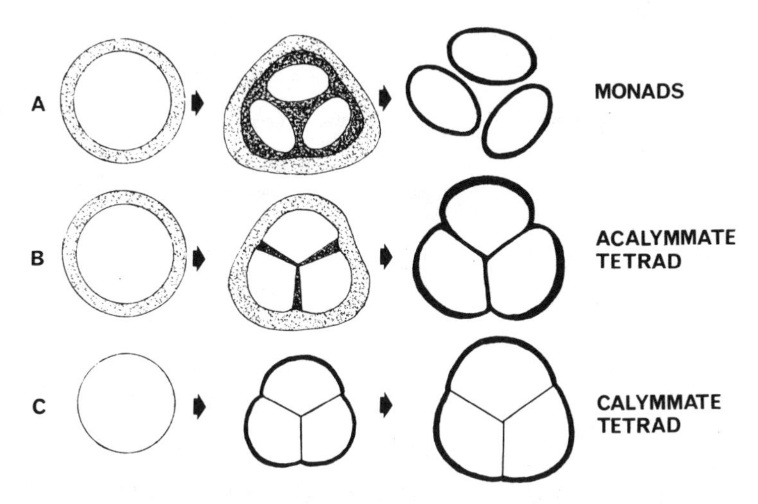

Figure 4.2. Origin of pollen dispersal units. Developmental processes in the formation of monads (A), acalymmate tetrads (B) and calymmate tetrads (C). In A and B the pollen mother cell secrets a common callose wall (light stipple) which surrounds the four microspores after meiosis. In A continued callose deposition forms individual callose walls (heavy stipple) which separate the microspores. Individual callose walls are reduced in B and absent in C leading to the development of partially (B) or completely (C) fused exines.

separate during cytokinesis, callose synthesis continues and each forms a callosic "individual special cell wall" (fig. 4.2A). Subsequently a primexine is formed between the plasma membrane and the individual callose wall and this acts as a template for the deposition of sporopollenin (Waterkeyn 1962; Heslop-Harrison 1966; Rowley and Skvarla 1975; Knox 1984; Barnes and Blackmore 1986).

Pollen in permanent tetrads arises either through complete loss, or a reduction in rate or duration, of callose synthesis by the mother cells and meiotic products (figs 4.2 and 4.3). Such grains may be viewed as paedomorphic in the sense of Alberch *et al.*(1979) with respect to this feature. The degree of cohesion between tetrad members varies considerably between taxa and on the basis of such differences two principal kinds of tetrad were recognized by Van Campo and Guinet (1961). They applied the term calymmate to tetrads in which the ectexines of the tetrad members are continuously fused while tetrads united only by partial, and sometimes extremely tenuous, fusion were termed acalymmate. Although

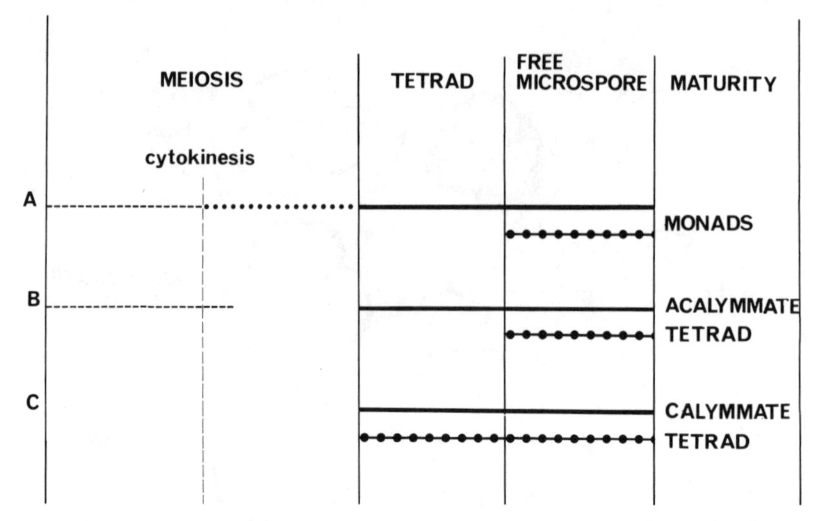

Figure 4.3. Duration and relative timing of callose and sporopollenin synthesis. Synthetic activity in the ontogeny of monads (A), acalymmate tetrads (B) and calymmate tetrads (C). In A and B a common callose wall (dashes) is deposited by the pollen mother cell during meiosis. In A callose deposition continues after cytokinesis to form individual callose walls (dots). Sporopollenin precursors of microsporal origin (solid line) may be produced in A, B, and C from the start of the tetrad stage. In A and B sporopollenin precursors of tapetal origin (line with dots) are prevented from reaching the microspores until after the dissolution of the callose walls at the end of the tetrad stage.

there are clear examples of both calymmate tetrads (certain Ericaceae, Gentianaceae, Hippocrataceae, and Nymphaeaceae Roland 1965), and acalymmate tetrads (Oenotheraceae, Mimosaceae, Winteraceae, Roland 1971), these two conditions intergrade and every intermediate degree of ectexine fusion may be encountered. Besides fusion of the ectexine Skvarla et al. (1976) and Takahashi and Sohma (1984) have emphasised that tetrad members may sometimes have narrow connections between the endexine or the intines.

If callose synthesis is entirely lacking, as reported for certain Juncaceae and Cyperaceae (Meyer and Yaroshevskaya 1976), the first material deposited forms a single continuous exine layer, termed sporocine, around the daughter cells and the resulting tetrads are calymmate. Callose synthesis is restricted, rather than completely absent during the formation of calymmate tetrads in other plants (fig. 4.2B). In Epacridaceae (Ford 1971, McGlone 1978) a common special cell wall is generated at cytokinesis but callose synthesis is then arrested and no individual special cell walls form. Primexine deposited around the individual microspores is not separated by callose. The developing pollen walls therefore fuse and the resultant tetrads are calymmate. Acalymmate tetrads occur when callose synthesis is arrested at a later stage so that a thin individual special cell wall covers the proximal faces of the daughter cells (fig. 2C) and thus provides some separation of primexine deposition (e.g., Monimiaceae, Sampson 1977).

In summary, the cohesion of pollen in permanent tetrads is a direct result of reduced callose synthesis (Figs 4.2–3). Although these changes in early ontogeny have little major effect on subsequent developmental processes, they do result in minor modifications of subsequent wall deposition. Waterkeyn (1962), Dunbar (1973), and others have suggested that the presence of a special cell wall is necessary for normal organization of the primexine and later the exine. Consequently, in pollen with permanent tetrads loss or reduction of callose on the proximal walls of the microspores results in modified ectexine development. Ford (1971) and Sampson (1977) have shown that the ectexine of proximal walls is thinner and related this to reduced primexine deposition. Thus, an early deviation from the normal ontogeny has both major immediate and minor delayed consequences.

In certain groups where permanent tetrads occur pollen dispersed as monads occurs as a derived condition (Ericales, Warner 1984, Erdtman 1952) but this reversal may be achieved by different means. In many taxa it seems likely that callose synthesis is simply extended, until individual special cell walls are formed around each microspore, but in the Cyperaceae and in *Styphelia* three of the microspores in each tetrad abort, resulting in "pseudomonads" (Erdtman 1952). In *Styphelia* Smith White (1959) demonstrated that three nuclei degenerate after meiosis whereas Dunbar (1973) and Strandhede (1973) have shown that abortion occurs much later in *Eleocharis,* after formation of ectexine is almost complete but before a phase of rapid size expansion.

ECTEXINE AND ENDEXINE

Sporoderm layers may be distinguished by either morphological (Erdtman 1948), or chemical criteria detectable with differential staining (Faegri and Iversen 1975). The morphological system is useful for descriptive palynology, particularly at the level of optical microscopy, while staining properties observed using transmission electron microscopy have been widely used as important characters in systematic investigations of such diverse groups as the Compositae (Skvarla and Larson 1965; Skvarla and Turner 1966; Skvarla et al. 1978) and Leguminosae (Ferguson and Strachan 1982; Ferguson and Skvarla 1981, 1983). Conventionally prepared ultrathin sections examined in the transmission electron microscope are stained with uranyl acetate and lead citrate. In mature pollen grains ectexine is more darkly stained (i.e., more electron dense) than endexine. In optical microscopy basic fuchsin stains ectexine dark red and endexine only lightly.

One unresolved question concerning exine stratification is the homology of the endexine in extant and fossil seed plants (fig. 4.4). This problem is complicated by the fact that routine laboratory preparations of pollen by acetolysis (Erdtman 1960) removes the intine. Since the intine may be interleaved with the endexine and contribute to its physical support, its absence in acetolyzed or fossil material may hinder correct interpretation of the presence or extent of the endexine. Some palynologists regard the endexine of angiosperms as homologous (fig. 2.4B) with that of gymnosperms (Guédès 1982) while others argue that the endexine of angiosperms is a new (fig. 2.4A) and different layer (Doyle et al. 1975; Zavada 1984). In the latter view the foot layer of angiosperms is considered to form on white-line centered lamellae prior to the dissolution of the callose wall and this is interpreted as indicating homology with the gymnosperm endexine. Kress and Stone (1982) also considered the timing of callose dissolution important in relation to the distinction between the ectexinous foot layer and the endexine; the former is considered to develop in the tetrad stage and the latter in the free microspore stage (see fig. 4.1). However, this temporal separation of endexine and ectexine deposition with respect to callose wall dissolution is not universal and, for example, in certain Compositae endexine deposition commences before the end of the tetrad period (Barnes and Blackmore 1986).

In our view the timing of the callose wall's dissolution is probably irrelevent to the distinction between ectexine and endexine because, as the example of permanent tetrads shows, callose synthesis and other events in sporoderm ontogeny may vary considerably in their relative times of initiation and in duration. More of the exine is deposited during the tetrad stage in gymnosperms and this may simply reflect the lesser

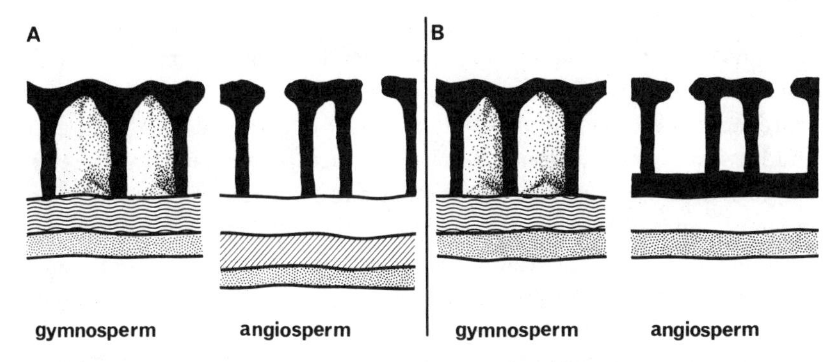

Figure 4.4. Two interpretations of sporoderm homology in spermatophytes. In one interpretation (A) the endexine of gymnosperms (wavy lines) is considered homologous to the foot layer of angiosperms (unshaded) while the endexine of angiosperms (diagonal shading) is thought to be a new layer. The ectexine (black) and intine (dotted) are considered homologous. In a second interpretation (B) the endexinous layers are considered homologous although that of angiosperms is unshaded to indicate the absence of lamellations at maturity. See text for further details.

contribution of their tapetum to the production of sporopollenin and other substances, and the greater degree to which elaborate morphogenetic control mechanisms have been developed in the comparatively derived angiosperms (Barnes and Blackmore 1986).

The initiation of ectexine development always precedes that of the endexine (Godwin et al. 1967, Faegri and Iversen 1975) and generally begins with sporopollenin accumulation within the primexine. Subsequently, sporopollenin of tapetal origin is usually added to the existing structure or may form additional layers. For example, the supratectal ornamentation in certain *Saxifraga* species (Hideux and Abadie 1981, 1985, 1986) and the foot layer of Compositae, and certain other groups, (Horner and Pearson 1978; Blackmore and Barnes 1985) are not formed on primexine. All sporopollenous wall components with a primexinous origin or which are added to the outer surface of layers derived from primexine are undoubtedly ectexinous. Structures that are added to the interior of the primexine derived wall components are more problematic. If these are deposited on circumferentially disposed white-line centered lamellae they are normally considered endexinous and thought to possess different staining properties. However, in a few instances radially oriented white-line centered lamellae have been implicated in ectexine deposition (Dickinson 1976; Sampson 1977; Rowley and Dahl 1977), and Dickinson and Heslop-Harrison (1968) considered ectexine and endexine to share a common mode of deposition since both involved lamellae. Faegri and Iversen (1975:33) doubted that the structures associated with ectexine formation were, in fact, membranes and pointed out that they are of different dimensions from those on which the endexine forms. Dickinson and Sheldon (1986) have now demonstrated that the apparent radial lamellae are not membranous. A further complication is that instances are known in which endexine is formed without the participation of such lamellae as in *Passiflora* (Larson 1966), *Helleborus* (Echlin and Godwin 1969) and *Austrobaileya* (Zavada 1984). The two exine layers therefore cannot be distinguished simply by timing of deposition, mode of deposition, or by their staining properties.

Nevertheless, ontogenetic criteria do provide the best available guide to distinguishing between ectexine and endexine. Ectexine may form in any of three distinct ways; on receptive sites within a primexinous matrix, on radially arranged lamellate structures that give rise to columellae, or by direct deposition from the tapetum onto pre-existing exinous surfaces. Endexine is formed either on circumferentially disposed white-line centered lamellae or in isolated instances, by direct accumulation of amorphous or granular material between plasma membrane and ectexine. In angiosperms white-line centered lamellae are not visible at maturity, using conventional preparative techniques, except in the vicinity of apertures. In gymnosperms such lamellae are not confined to apertures at maturity. Zavada (1984) suggested that the presence of lamellations at maturity in gymnosperm endexine demonstrates its homology with the angiosperm foot layer. However, in angiosperms the foot layer, like the endexine, lacks such lamellations at maturity and, in our interpretation and that of Guédès (1982), a true foot layer is not lamellated, even during development.

Taxa in which lamellations are present immediately below the columellate or granular ectexine are considered by us to lack a foot layer. The absence in mature angiosperm pollen of endexine lamellations does not alter the fact that the layer is initially deposited on white-line centered lamellae in precisely the same way as the endexine of gymnosperms (Audran 1981; Dickinson and Bell 1970; Dickinson 1976), the inner exospore of pteridophytes (Lugardon 1976) and layers constituting most of the wall in bryophyte spores (Neidhart 1979). In algae which posses sporopollenin it is also deposited on white-line centered lamellations (Atkinson et al. 1972). In angiosperms the endexine is probably a primitive wall layer (Faegri and Iversen 1975) and in derived taxa only vestiges of endexine are retained at the apertures (e.g. grasses, Rowley 1964; Skvarla and Larson 1966; Betulaceae, Dunbar, and Rowley 1984).

In contrast the ectexine is a derived wall layer with a mode of deposition that is simple in lower plants and more complex in higher plants. The increasing complexity of ectexine structure and ontogeny in higher plants has involved the origin of novel morphogenetic control mechanisms. For example, an elaborate primexine with delayed differentiation may give rise to a structurally complex ectexine or the callose wall of the tetrad may be involved in determining exine patterning as in the derived pantoporate pollen of *Ipomoea* (Waterkeyn and Bienfait, 1970) and echinolophate pollen of *Scorzonera* (Barnes and Blackmore 1986).

REDUCED EXINES

Several unrelated groups of flowering plants have vestigial or highly reduced exines (see Kress 1986). In all cases where the exine is greatly reduced the intine is thickened and unusually elaborate, possessing characteristics that occur only in the apertures of most pollen grains (Rowley and Skvarla 1970). Ontogenetic details have been established for two monocotyledonous groups, marine angiosperms belonging to the Hydrocharitaceae and Cymodoceaceae (Pettitt et al. 1978, 1981; Pettitt 1981), and members of the Zingiberales (Rowley and Skvarla 1970, 1975; Stone et al. 1979, 1981; Kress and Stone 1982). Species of the dicotyledonous Lauraceae and Hernandiaceae have been investigated ultrastructurally by Kubitzki (1981) and Hesse and Kubitzki (1983) and ontogenetically by Stone (1985). In almost every case primexine is deposited, but the later stages of exine differentiation are deleted although intine development continues. Intine formation may be variously modified, to produce, for example, the distinctly channeled intines of the Lauraceae. The differences that arise in the extent of exine present depend on the point at which normal exine development is interrupted. From these studies a very simple model (fig. 4.5) can be used to summarize the relationships between ontogeny and morphology.

Pollen grains that completely lack exines are known only from marine

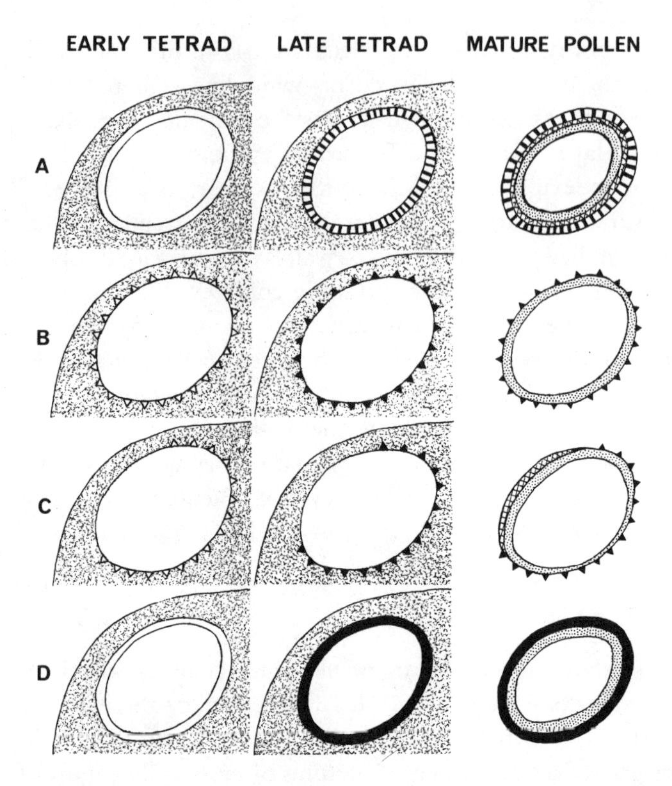

Figure 4.5. Origin of typical and exineless pollen grains. In a typical angiosperm pollen grain (A) a primexine (unshaded) is deposited within the callose wall (stippled) during the early tetrad stage. Later, with the addition of sporopollenin, primexine differentiates to form ectexine (black) and endexine (cross hatched) and intine (dotted) are deposited. In *Canna* (B), primexine is limited to spinules, the only sites at which sporopollenin is subsequently added. A thickened intine develops. In *Heliconia* (C), early stages resemble *Canna* but primexine is absent from the distal surface where a thin endexine is later deposited. In *Tapeinochilos* (D), an extensive primexine forms but no phase of differentiation occurs during sporopollenin accretion; endexine is absent. The diagrams are based on interpretations of the published works cited in the text.

angiosperms. Pettitt et al. (1981) demonstrated that no primexine is formed by microspores when in tetrads and that the major part of the wall of the mature pollen grains is composed of intine. In the pollen of *Canna generalis* the exine is present only as scattered spinules (fig. 4.5B).

In one of the clearest expositions on sporopollenin receptive sites in primexines Rowley and Skvarla (1975) demonstrated that these correspond to particular regions within the restricted primexine generated during the terad stage. Sporopollenin accumulation is limited to a meshwork of polysaccharides and proteins identified by Rowley and Skvarla as the receptive area of the primexine. No other form of sporopollenin accretion occurs, for example on white-line centred lamellae, and the next wall layer is an elaborately channelled bi-layered intine which begins to form after microspore mitosis.

Sporoderm development in *Heliconia* (fig. 4.5C) follows a similar course to that of *Canna* except that exine development is much more extensive on the proximal faces of developing microspores (Stone et al. 1979, Kress and Stone 1982). The distal surfaces are somewhat similar to *Canna* pollen in possessing scattered exinous spinules but there is also limited sporopollenin accretion on white-line centered lamellae. The more continuous exine of the proximal face is formed mainly on white-line centered lamellations and is completed before the end of the tetrad phase. Stone and his co-workers considered this layer to be a limited ectexinous foot layer (or nexine 1) and pointed out that although the tapetum is active in the free spore stage little or no sporopollenous material is added to the walls by the tapetum during this period. As in *Canna* the intine is multilayered, and this is most distinct on the distal (apertural) surface where the least exine is present.

In *Tapeinochilos* Stone et al. (1981) showed that a thick undifferentiated primexine is formed during the tetrad stage (fig. 4.5D), except in areas destined to become apertures. The primexine is later reinforced by the addition of sporopollenin. No further sporopollenin is added during the free spore stage. *Tapeinochilos* pollen grains were considered by Stone et al. (1981) to provide an example of neoteny since they are protected at maturity by primexine-like sporoderms of "juvenile" origin. A thin endexine is deposited on a few white-line centered lamellations at the distal pole. Intine development is reduced compared to the pollen grains of most other Zingiberales and the intine is distinctly bi-layered only at the apertures.

In the dicotyledonous Laurales, pollen grains with reduced exine resembling those of the Zingiberales are formed in a similar way. Hesse and Kubitzki (1983) remarked that channeled intine and reduced exine were apparently limited to inaperturate pollen types. While this might be the case in the Laurales it seems unlikely that there is a direct correlation between reduced exine and absence of apertures. If the reduction of primexine is so extensive that apertures are no longer morphologically defined, as in *Canna,* then inaperturate or omniaperturate pollen would result. In *Tapeinochilos* the primexine is deposited differentially in a manner that gives rise to apertures but does not undergo differentiation into columellate or granular forms. It should also be noted that the presence of thick, channelled intine is not restricted to pollen grains with reduced exines although it may indeed be associated with modified ontogeny. In *Thunbergia* (Acanthaceae), for example, the intine is thick and channelled but the pollen grains are spiraperturate with granular ectexine (Raj 1961; Furness in press).

APERTURE FORMATION

Apertures are defined as thin, preformed regions of the exine through which germination may occur (Erdtman 1952; Walker and Doyle 1975). Although there is considerable variation in aperture type, number, and position in flowering plants (Erdtman 1969) in the context of tracheophytes as a whole there are two basic apertural types. In pteridophytes and a few primitive seed plants there is a single proximal monolete or trilete aperture, whereas in most seed plants there is a single distal or three equatorial apertures (or variations derived from these basic types). Proximal apertures occur in the mosses, thought to be the sister group to the tracheophytes (Mishler and Churchill 1984) and therefore recognition of the transformation proximal to distal/equatorial apertures is clear. Ontogenetic studies clearly demonstrate that the mode of formation of proximal and distal/equatorial apertures is quite distinct. A proximal trilete or monolete suture arises simply as a scar resulting directly from microspore contact in a tetrahedral or tetragonal tetrad. In contrast distal or equatorial apertures are formed by the intervention of some kind of organelle shield, typically endoplasmic reticulum, during wall development that prevents primexine deposition across regions of the plasma membrane destined to become apertures. The single trend of aperture evolution (proximal to distal/equatorial) is therefore resolved as two separate characters, presence or absence of proximal tetrad scar and presence or absence of distal/equatorial apertures. This is further supported by the occurrence of certain fossil pollen grains that exhibit both proximal monolete or trilete marks and a distal aperture (e.g., *Potoniesporites;* the pollen of the early conifer *Lebachia,* Mapes and Rothwell 1984; Scott and Chaloner 1983). Within tracheophytes as a whole then there appears to have been the loss of the developmental processes responsible for proximal "aperture" formation, and the addition of an organelle shield and patterned plasma membrane during development that results in the formation of distal/equatorial apertures characteristic of most seed plants.

SACCUS AND CAVEA FORMATION

In seed plants as a whole the presence of sacci in conifers, cordaites, and several other groups (Crane 1985a) is a derived feature relative to the non-saccate pollen (prepollen; Chaloner 1970; Schopf 1938) of primitive seed plants such as *Lyginopteris* (Crane 1985a). Observations on the ontogeny of saccate pollen grains of extant *Pinus* (Dickinson and Bell 1970; Dickinson 1976) show that sacci are formed by the incorporation of polysaccharide into specific sites on the distal wall following differentiation of the primexine but prior to the dissolution of the callose special cell wall. Later in development, just before dehiscence of the microsporangia, this polysaccharide expands dramatically to inflate the sacci. The polysaccharide matrix is subsequently reabsorbed leaving the sacci as air filled cavities in the wall. The presence of sacci is therefore dependent on two additions to the developmental process seen in most gymnosperms, intercalation of polysaccharide deposition between primexine differentiation and exine deposition, and the incorporation of a polysaccharide expansion phase just before microsporangial dehiscence.

The mechanism of saccus formation is similar to the mode of cavea formation in pollen grains of some Compositae (Blackmore et al. 1984) although phylogenetically these two mechanisms have clearly evolved independantly. The caveae in the pollen grains of certain Compositae are air filled spaces in the ectexine between the foot layer and the columellae. These caveae are formed by the absence during development of sporopollenin receptive sites in the microfibrillar spacer layer (Horner and Pearson 1978) which separates the outer part of the wall (formed on primexine) from the inner (formed on lamellae or by tapetal deposition). The spacer layer may well be a modified component of the predominantly polysaccharide primexine but since it does not accumulate sporopollenin it is later reabsorbed leaving air or pollenkitt filled cavities in the wall. In contrast to the situation in saccate grains development of the caveae involves the absence of receptive sites for sporopollenin deposition in specific regions of the primexine.

DISCUSSION

The examples reviewed above demonstrate that in pollen and spore development, just as in the ontogeny of other organs and organisms (Gould 1977), relatively minor perturbations may have pronounced phenotypic effects. These kinds of developmental changes are of major evolutionary significance and almost certainly account for much of the enormous diversity that has been so extensively documented in the palynological literature.

Some of the ontogenetic changes considered above appear to be straightforward deletions or additions to a basic developmental program. Additions, for example, would include the origin of an organelle shield and patterned plasma membrane permitting the development of distal/equatorial apertures in most seed plants, and the addition of tapetally derived proteins to the pollen wall during the final stages of maturation of angiosperm pollen. Examples of deletions would include loss of individual callose wall synthesis leading to calymmate permanent tetrads in Ericaceae and the loss of primexine differentiation during the development of *Tapeinochilos* pollen. However, there is enormous variation, both interspecific and intraspecific, in the extent to which the developmental changes are expressed. For example distal/equatorial apertures vary markedly in the extent to which they are clearly defined ranging from indistinct in *Datura* and *Scopolia* to distinct in most other Solanaceae (Punt and Monna-Brands 1977). Clearly defined additions and deletions are rare and it is more realistic to view ontogenetic modifications as changes in the relative timing and rate of several basic developmental processes that are operating simultaneously (Gould 1977; Alberch et al. 1982) rather than as simplistic all or nothing effects. Stone et al. (1981) for example interpret the exine structure of *Tapeinochilos* as neoteonous with respect to that in other orders of monocotyledons and the non-saccate grains of certain conifers, e.g. *Juniperus,* could be regarded as progenetic with respect to those of cordaites and cladistically more primitive conifers.

Studies of heterochrony in pollen development are certain to be of considerable future interest in palynology, but because such studies are necessarily comparative they are "fundamentally dependent on hypotheses of phylogeny" determined by outgroup analysis in preference to ontogeny (Fink 1982:254).

Ontogeny and phylogenetic systematics

Studies of ontogeny have three applications in phylogenetic systematics: (i) they provide additional characters of value in constructing classifications, (ii) they increase our understanding of characters and therefore provide additional information-that may corroborate or refute hypotheses of homology based on examination of the mature condition, and (iii) they provide direct evidence from which to assess the polarity of character transformations (Kluge 1985).

In general the current paucity of comparative ontogenetic data has so far limited the direct application of developmental mechanisms as characters in palynology. Sporangial development is an important criterion in distinguishing between eusporangiate and leptosporangiate ferns and occurrence of successive or simultaneous microspore cytokinesis has been used in assessing relationships between higher taxa of flowering plants (Dahlgren and Bremer 1985).

Similarly, developmental data have seen little application in testing hypotheses of homology based only on mature pollen. Wodehouse (1935) and Jonker (1974), suggested that the trichotomosulcate aperture of *Schisandra* was homologous to the proximal trilete mark of fern spores. Examinations of *Schisandra* pollen at tetrad stage and in occasional permanent tetrads (Erdtman 1971:257; Walker 1974; Huynh 1976b; Praglowski 1976) demonstrated that the aperture was distal rather than proximal and thus the potential homology was refuted by the positional criterion detected developmentally.

However, developmental data are not always unambiguous. It has been inferred that the endexine of gymnosperms and angiosperms is nonhomologous based primarily on different mature structure and the timing of endexine deposition (Zavada 1984). Gymnosperm endexine being laminated at maturity and deposited during the tetrad phase, and angiosperm endexine being nonlaminated (except adjacent to apertures; Doyle et al. 1975) and deposited during the free spore stage. In view of the possibility of widespread heterochrony between different processes in pollen development criteria of homology based on timing should be treated cautiously. Angiosperm and gymnosperm endexines are both deposited on white-line centered lamellae and in our view are almost certainly homologous. The absence of striations in the mature endexine of angiosperms may be attributable to the subsequent incorporation of tapetally derived sporopollenin. In gymnosperms the incorporation of tapetally derived sporopollenin into the gametophytically derived wall is less pronounced.

The third, and most controversial application of ontogenetic data in phylogenetic systematics concerns its rôle in determining character polarity—that is, determining the direction of change in character phylogenies. Opinions are divided between those who regard ontogenetic criteria as useful, but of secondary importance to outgroup comparison (Brooks and Wiley 1985; Kluge 1985), and those who regard ontogeny as the final arbiter in assessing the polarity of homologies (Nelson 1985; Patterson 1982). According to Nelson and Platnick (1981:37) "features that appear early in development can be considered primitive relative to the modifications of those features that appear later in development." This is proposed as a law potentially open to falsification (see Kluge 1985:18–19; Voorzanger and van der Steen 1982; and reply by Nelson 1985: 35–36) based on the principle that "ontogeny is orderly (i.e., character relations are universal and contradictions do not arise)" (Nelson 1985:36). The controversy over the ontogenetic criterion is intimately linked with the active debate over the extent to which "ontogeny provides strictly empirical evidence of pattern" (Kluge 1985:14), and the apparent conflict between the two schools labeled "pattern" and "phylogenetic" cladistics. According to Kluge (1985) ontogeny does not always provide unambiguous evidence of character polarity, and Brooks and Wiley (1985:1) state that "direct observation of ontogeny does not resolve any cases that outgroup comparison fails to resolve, and outgroup comparison does resolve some cases where direct observation fails."

In this paper it is not our intention to argue for, or against, these two alternative views at a philosophical level but merely to provide our assessment of the practical utility of ontogenetic criteria on the basis of current knowledge of pollen and spore development and other botanical systems.

We consider two basic issues; is pollen and spore ontogeny orderly, and what is the relative practical utility of the ontogenetic criterion versus outgroup comparison in comparative palynology?

Is pollen and spore ontogeny orderly?

The development of a typical pollen wall involves sequential deposition of callose, sporopollenous exine, and pecto-cellulosic intine. Currently available data on tetrad formation and exine deposition demonstrate that these three basic phases may be variously modified, even to the extent of complete, or almost complete, deletion of one or more phases. However, in all the examples of which we are aware the ontogenetic sequence of initiation of the three phases is always the same. Intine deposition never precedes exine deposition which in turn never precedes callose deposition. Relative diminution or expansion of the various phases has been documented, but the relative positions of those phases are never reversed. Similarly, in all cases in which there is a separation between primexine deposition and deposition of endexine on lamellae the initiation of the former always precedes that of the latter. However it should be recognized that this orderliness does not extend to order of completion of the various processes. For example, the onset of endexine deposition and callose dissolution is reversed in some angiosperms relative to the situation in gymnosperms.

In certain cases, then, the ontogenetic sequence appears to be completely orderly but only with respect to the initiation of processes, rather than orderly transformation of one structure into another. Each process is responsible for the production of a certain kind of wall but there is no transformational relationship between the different kinds of walls that provides direct evidence of character polarity. Indeed, we know of very few situations in palynology in which such a direct transformational relationship exists between two homologies. The best documented example is the ornamentation of *Saxifraga* pollen (Hideux and Abadie 1985, 1986) in which relatively derived species pass through stages in development in which their ornamentation corresponds to that in pollen of relatively generalized species. Thus, although "all characters (Synapomorphies) are potentially understandable as ontogenetic transformations" (Nelson 1985:42) in practice, at least in palynology, they are only understandable in terms of transformations in processes rather than transformations in structures.

The ontogenetic criterion versus outgroup comparison in comparative palynology

There are many examples of palynological character transformations that are unresolved by ontogeny but resolved by outgroup comparison. The transition from a distal monosulcate aperture to three equatorially placed apertures within flowering plants is one example. The triaperturate (or triaperturate derived) condition occurs only in the "higher dicots" (subclasses Hamamelidae, Caryophyllidae, Rosidae, Dilleniidae, Asteridae of Cronquist 1981), while a distal monosulcate (or monosulcate derived) aperture is found in monocotyledons and magnoliid dicots. Only the monosulcate condition occurs in gymnosperms. By outgroup comparison the triaperturate condition is interpreted as apomorphic, the monosulcate condition as plesiomorphic—and this is also consistent with palaeobotanical data. Monosulcate angiosperm pollen appears in the fossil record earlier than triaperturate angiosperm pollen (Muller 1970, Doyle and Hickey 1976). Ontogenetic data is of no value in determining the polarity of this character transformation, and triaperturate grains do not pass through a monosulcate phase in their development. However, as far as it is currently understood, the developmental basis of aperture formation is identical in both cases. The triaperturate condition merely represents an increase in number and relative position, probably as a result of a shift from a tetragonal to a tetrahedral tetrad configuration.

Although pollen grains of different plants may more properly be regarded as homonomous rather than strictly homologous (Patterson 1982, Wiley 1981) at the level of individual pollen grains the transition to monosulcate to triaperturate constitutes a transition from special homology to serial homology (Wiley 1981:11). The ontogenetic criterion in this case fails, because developmentally the three equatorial apertures and the single distal aperture form in exactly the same way, merely by the duplication and repositioning of a single developmental process. Following the argument adopted by Patterson (1982:54) we could conclude that three equatorial apertures is a potential synapomorphy relative to absence of the same character, and that a monosulcate aperture is a potential synapomorphy relative to the absence of such an aperture. However, the ontogenetic criterion cannot resolve how the two characters, monosulcate and triaperturate, stand in relation to each other. The inability of the ontogenetic criterion to determine polarity in this kind of character transformation may seriously limit its practical utility in systematic botany.

CONCLUSIONS

The study of pollen ontogeny clearly has a valuable role to play in providing additional characters for constructing classifications, and increasing our understanding of characters in such a way as to provide a similarity test of hypotheses of homology based on the mature condition (Patterson 1982). However, the role of ontogeny in determining the polarity of palynological characters is more problematic. Even though there are many aspects of pollen and spore ontogeny that are not understood, it is nevertheless a relatively simple process with little inherent pattern. Pollen ontogeny is not a highly structured, contingent process in which the modification of one structure into another can very frequently be observed. It is instead continuous and itself controlled by the duration and rate of predominantly intracellular processes operating at the level of organelles and molecules.

Pollen and spore ontogeny is essentially a process of cell differentiation, comparable to the development of a tracheid or a unicellular trichome, and it is reasonable to question whether generalizations based on such systems are applicable to other levels of plant morphogenesis involving tissue or organ differentiation. To a very large extent the ontogeny of higher plants involves the repeated formation of discrete morphological units such as leaves, flowers, hairs, or pollen grains. This process is distinct from the more integrated "unitary" development of higher animals (White 1979, Schnaar 1985) and may also account for the perceived high degree of parallelism in many botanical cladograms. Much of plant diversity is therefore based on duplicated morphological units that have discrete, rather simple, noncontingent, developmental histories, and that are arranged in a variety of different combinations. In this context, the kind of ontogenetic processes that control pollen and spore morphology may not differ fundamentally from those that govern other aspects of plant development. It is also possible that mechanisms analogous to the oligosaccharide control of intercellular morphogenesis (Albersheim and Darvill 1985; Tran Thanh Van et al., 1985) occur in association with the polysaccharide coat, or glycocalyx (Bennett 1963, 1969) on the surface of developing pollen and spores (Rowley 1978, 1981).

In some areas of botany there are well known examples in which transitions between character states do appear to occur in the course of ontogeny. For example, the unusual elongated stomata of some grasses pass through a stage in their early ontogeny in which they are "D-shaped" like those of most other embryophytes (Flint and Moreland 1946). Dictyostelic and solenostelic pteridophytes pass through a stage in their early ontogeny in which they are protostelic, apparently as a result of the initially small size of the apical meristem. In these cases polarity determinations based solely on ontogeny, or on outgroup comparison (or even the fossil record) yield the same result. However, Patterson (1982) and others have argued that the ontogenetic criterion is theoretically preferable to outgroup comparison because it does not require any prior knowledge of relationship. This position cannot be refuted either with our data, or on theoretical grounds, but it is nevertheless a position that confronts practical difficulties. Studies of morphogenesis remain to be fully exploited in plant systematics, but the modular structure and developmental mechanisms that characterize higher plants necessarily reduce potential applications of the simple ontogenetic criterion. Outgroup comparison has been viewed merely as "a method of ordering homologies of unknown polarity, so that they are congruent with polarities determined by ontogeny" Patterson 1982:54–55). If outgroup comparison is strictly relegated to this secondary role then relationships among higher plants will inevitably remain largely unresolved. Whatever its theoretical status, outgroup comparison seems certain to persist as the primary practical means of determining character polarity in plant systematics.

ACKNOWLEDGMENTS

We thank C.J. Humphries for inviting us to participate in this symposium and for his constructive comments on the manuscript. We are grateful to P.J. Stafford for illustrating the text and to S.H. Barnes, H.G. Dickinson and J.R. Rowley for many discussions on sporoderm ontogeny.

REFERENCES

Abadie, M. and M. Hideux, 1979. L'anthère de *Saxifraga cymbalaria* L. ssp. *huetiana* (Boiss.)Engl. and Irmsch. en microscopie électronique (M.E.B. et M.E.T.). 2. Ontogenèse du sporoderm. *Ann. Sci. Nat. Bot.* 1: 237–281.

Alberch, P. 1982. Developmental constraints in evolutionary process. In J.T. Bonner, ed. *Evolution and Development* pp 313–332. Berlin: Springer-Verlag. Alberch, P., S.J. Gould, G.F. Oster, and D.B. Wake. 1979. Size and shape in ontogeny and phylogeny. *Paleobiology* 5: 296–317.

Albersheim, P, and A.G. Darvill. 1985. Oligosaccharins. *Scientific American* 253: 44–50

Atkinson, A.W., B.E.S. Gunning, and P.C.L. John, 1972. Sporopollenin in the cell wall of *Chlorella* and other algae: ultrastructure, chemistry and incorporation of C14–Acetate studied in synchronous culture. *Planta* 107: 1–32.

Audran, J.C. 1981. Pollen and tapetum development in *Ceratozamia mexicana* (Cycadaceae): sporal origin of the exinic sporopollenin in cycads. *Rev. Palaeobot. Palynol.* 33: 315–346.

Barnes, S.H. and S. Blackmore 1984a. Freeze fracture and cytoplasmic maceration in botanical scanning electron microscopy. *J. Microsc.* 136:3–4.

Barnes, S.H. and S. Blackmore. 1984b. Scanning electron microscopy of chloroplast ultrastructure. *Micron and Microscopica Acta* 15: 187–194.

Barnes, S.H. and S. Blackmore. 1986. Some functional features during pollen development. In S. Blackmore and I.K. Ferguson (Eds), *Pollen and Spores: Form and Function.* London: Academic Press.

Bennett, H.S. 1963. Morphhological aspects of extracellular polysaccharides. *J. Histochem. Cytochem.* 11: 14–23.

Bennett, H.S. 1969. The cell surface: components and configurations. In A. Lima de Faria (Ed.), *Handbook of Molecular Cytology.* pp 1261–1293. Amsterdam and London: North Holland Publishing Co.

Blackmore, S. 1982. Palynology of subtribe Scorzonerinae (Compositae: Lactuceae) and its taxonomic implications. *Grana* 21: 149–160.

Blackmore, S. 1984. Pollen features and plant systematics. In V.H. Heywood and D.M. Moore (eds), *Current Concepts in Plant Taxonomy,* pp. 135–154. London and Orlando: Academic Press.

Blackmore, S. and M.J. Cannon, 1983. Palynology and systematics of Morinaceae. *Rev. Palaeobot. Palynol.* 40: 207–226.

Blackmore, S., H.A.M. van Helvoort, and W. Punt. 1984. On the terminology, origins and functions of caveate pollen in Compositae. *Rev. Palaeobot. Palynol.* 43: 2933–01.

Blackmore, S. and S.H. Barnes, 1985. *Cosmos* pollen ontogeny, a scanning electron microscope study. *Protoplasma* 121: 91–99.

Blackmore, S. and S.H. Barnes, 1986. In press Harmomegathic mechanisms in pollen grains. In S. Blackmore and I.K. Ferguson, eds. *Pollen and Spores: Form and Function.* London: Academic Press.

Bolick, M.R. 1983. A cladistic analysis of the Ambrosiinae Less. and Engelmanniinae Stuessy. In V.A. Funk and N. Platnick (eds.), *Advances in Cladistics,* pp. 125–141. New York Botanical Garden.

Bremer, K. and Wanntorp, H.-E. 1978. Phylogenetic systematics in botany. *Taxon* 27: 317–329.

Brooks, D. R. and E.O. Wiley, 1985. Theories and methods in different approaches to phylogenetic systmatics. *Cladistics* 1: 1–11.

Chaloner, W.G. 1970. The evolution of miospore polarity. *Geoscience and Man* 1: 47–56.

Crane, P.R. 1985a. Phylogenetic analysis of seed plants and the origin of angiosprms. *Ann Mo. Bot. Gard.* 72: 716–793

Crane, P.R. 1985b. Phylogenetic relationships in seed plants. *Cladistics,* 1:329–348

Cronquist, A. 1981. *An integrated system of classification of flowering plants.* New York: Columbia University Press,

Cruden, R.W. 1977. Pollen ovule ratios: a conservative indicator of breeding systems in the flowering plants. *Evolution* 31: 32–46.

Dahlgren, R. and K. Bremer, 1985. Major clades of the Angiosperms. *Cladistics* 1: 349–368.

Dickinson, H.G. 1976. Common factors in exine deposition. In I.K. Ferguson and J. Muller (eds), *The Evolutionary Significance of the Exine,* pp. 67–89. London: Academic Press.

Dickinson, H.G. and Heslop-Harrison, J. 1968. Common mode of deposition for the sporopollenin of sexine and nexine. *Nature* 220: 926–927.

Dickinson, H.G. and P.R. Bell 1970. The development of the sacci during pollen formation in *Pinus banksiana. Grana* 10: 101–108.

Dickinson, H.G. and D. Lewis, 1973. The formation of the tryphine coating the pollen grains of *Raphanus,* and its properties relating to the self-incompatibility system. *Proc. R. Soc. Lond., B.,* 184: 149–165.

Dickinson, H.G. and J. Sheldon, 1986. The generation of patterning at the plasma membrane of the young microspore of *Lilium.* In S. Blackmore and I.K. Ferguson (Eds) (1986), *Pollen and Spores: Form and Function.* London: Academic Press.

Doyle, J.A., M. Van Campo and B. Lugardon, 1975. Observations on exine structure of *Eucommidites* and lower Cretaceous angiosperm pollen. *Pollen Spores* 17: 429–486.

Doyle, J.A. and L.J. Hickey, 1976. Pollen and leaves from the Mid-Cretaceous Potomac group and their bearing on early angiosperm evolution. In C.B. Beck (Ed.), *The Origin and Early Evolution of the Angiosperms,* pp. 139–206. New York: Columbia University Press.

Dunbar, A. 1973. Pollen development in the *Eleocharis palustris* group (Cyperaceae). *Bot. Not.* 126: 197–254.

Dunbar, A. and J.R. Rowley, 1984. *Betula* pollen development before and after dormancy: exine and intine. *Pollen Spores* 26: 299–338.

Echlin, P. and Godwin, H. 1969. The ultrastructure and ontogeny of pollen in *Helleborus foetidus* L. II. Pollen grain development through the callose special wall stage. *J. Cell Sci.* 3: 175–186.

Erdtman, G. 1945. Pollen morphology and plant taxonomy. V. On the occurence of tetrads and dyads. *Svensk Bot. Tidskr.* 29: 286–297.

Erdtman, G. 1948. Did dicotyledonous plants exist in early Jurassic times? *Geol. Foreningens Forhandlingar* 70: 265–271.

Erdtman, G. 1960. The acetolysis technique: a revised description. *Svensk Bot. Tidskr.* 54: 561–564.

Erdtman, G. 1952. *Pollen Morphology and Plant Taxonomy. Angiosperms.* Stockholm: Almqvist and Wiksell.

Erdtman, G. 1969. *Handbook of Palynology.* Copenhagen: Munksgaard,

Erdtman, G. 1971. *Pollen Morphology and Plant Taxonomy. Angiosperms.* 3d ed. New York: Hafner.

Erdtman, G. and P. Sorsa, 1971. *Pollen and Spore Morphology Plant Taxonomy: Pteridophyta.* Stockholm: Almqvist and Wiksell.

Faegri, K. and J. Iversen, 1975. *Textbook of Pollen Analysis.* 3rd ed. New York. Hafner.

Ferguson, I.K. and R. Strachan, 1982. Pollen morphology and taxonomy of the tribe Indigofereae (Leguminosae: Papilionoideae). *Pollen Spores* 24: 171–210.

Ferguson, I.K. and J.J. Skvarla, 1981. The pollen morphology of the subfamily Papilionoideae (Leguminoase). In R.M. Polhill and P.H. Raven, eds., *Advances in Legume Systematics,* pp. 859–896. Kew: Royal Botanic Gardens.

Ferguson, I.K. and J.J. Skvarla, 1983. The granular interstitium in the pollen of subfamily Papilionoideae (Leguminosae). *Amer. J. Bot.* 70: 1401–1408.

Fink, W. 1982. The conceptual relationship between ontogeny and phylogeny. *Paleobiology* 8: 254–264.

Flint, L.H. and C.F. Moreland, 1946. A study of the stomata in sugar cane. *Amer. J. Bot.* 33: 8–82.

Ford, J.H. 1971. Ultrastructural and chemical studies of pollen wall development in Epacridaceae. In J. Brooks, P.R. Grant, M. Muir, P. van Gijzel, and G. Shaw, eds. *Sporopollenin,* pp. 130–173. London and New York: Academic Press,

Funk, V.A. and T. Stuessy, 1978. Cladistics for the practicing plant taxonomist. *Syst. Bot.* 3: 159–178.

Furness, C.A. in press. A review of spiraperturate pollen. *Pollen Spores,*

Godwin, H., P. Echlin, and B. Chapman, 1967. The development of the pollen grain wall in *Ipomoea purpurea* (L.) Roth. *Rev. Palaeobot. Palynol.* 3: 181–195.

Gould, S.J. 1977. *Ontogeny and Phylogeny.* Cambridge Mass.: Harvard University Press.

Guédès, M. 1982. Exine stratification, ectexine structure and angiosperm evolution. *Grana* 21: 161–170.

Heslop-Harrison, J. 1963. Ultrastructural aspects of differentiation in sporogenous tissues. *Symp. Soc. exp. Biol.* 25: 277–300.
Heslop-Harrison, J. 1966. Cytoplasmic continuities during spore formation in flowering plants. *Endeavour* 25: 65–72.
Heslop-Harrison, J. 1968. The pollen grain wall. *Science* 161: 230–237.
Heslop-Harrison, J. 1976. The adaptive significance of the exine. In I.K. Ferguson and J. Muller (eds), *The Evolutionary Significance of the Exine.* pp. 27–37. London: Academic Press.
Heslop-Harrison, J. 1979. An interpretation of the hydrodynamics of pollen. *Amer. J. Bot.* 66: 77–743.
Hesse, M. 1978. Entwicklungsgeschichte und Ultrastruktur von Pollenkitt und Exine bei nahe verwandten entomophilen, anemophilen angiospermensippen, Ranunculaceae, Hamamelidaceae, Platanaceae und Fagaceae. *Pl. Syst. Evol.* 130: 13–42.
Hesse, M. and K. Kubitzki, 1983. The sporoderm ultrastructure in *Persea, Nectandra, Hernandia, Gormortega* and some other Lauralean genera. *Pl. Syst. Evol.* 141: 299–311.
Hideux, M. and M. Abadie, 1981. The anther of *Saxifraga cymbalaria* L. ssp. *heutiana* (Boiss.)Engl. and Irmsch.: a study by electron microscopy (S.E.M. and T.E.M.). *Annales de Sciences Naturelles, Botanique, Paris* 13: 27–37.
Hideux, M. and M. Abadie, 1985. Cytologie ultrastructurale de l'anthère de *Saxifraga.* I. Periode d'initiation des precurseurs des sporopolleniques au niveau des principaux types exiniques. *Can. J. Bot.* 63: 97–112.
Hideux, M. and M. Abadie, 1986. Ontogenetic constraints on function in pollen of some *Saxifraga* L. species. In S. Blackmore and I.K. Ferguson eds. *Pollen and Spores: Form and Function.* London: Academic Press,
Hill, C.R. and P.R. Crane, 1982. Evolutionary cladistics and the origin of angiosperms. In K.A. Joysey and A.E. Friday (eds), *Problems of Phylogenetic Reconstruction,* pp. 269–362. New York: Academic Press.
Horner, H.T. and C.B. Pearson, 1978. Pollen wall and aperture development in *Helianthus annuus* (Compositae: Heliantheae). *Amer. J. Bot.* 65: 293–309.
Humphries, C.J. 1979. A revision of the genus *Anacyclus* L. (Compositae: Anthemideae). *Bull. Br. Mus. nat. Hist. (Bot.)* 7: 83–142.
Humphries, C.J. and V.A. Funk, 1984. Cladistic methodology. In V.H. Heywood and D.M. Moore (eds), *Current Concepts in Plant Taxonomy,* pp. 323–362. London and Orlando: Academic Press.
Huynh, K.L. 1973. L'arrangement des spores dans le tétrade chez les Ptéridophytes. *Bot. Jahrb.* 93: 9–24.
Huynh, K.L. 1976a. Arrangement of some monosulcate, disulcate, trisulcate, dicolpate and tricolpate pollen types in the tetrads, and some aspects of evolution in the angiosperms. In: I.K. Ferguson and J. Muller (eds), *The Evolutionary Significance of the Exine* pp. 101–124. London: Academic Press.
Huynh, K.L. 1976b. L'arrangement du pollen du genre *Schisandra* (Schisandra-

ceae) et sa significance phylogénique chez les Angiosperms. *Bietr. Biol. Pflanzen.* 52: 227–253.
Jonker, F.P. 1974. Reflections on pollen evolution. *Adv. in Pollen Spore Res.* 1: 50–61.
Kenrick, J. and R.B. Knox, 1982. Function of the polyad in reproduction of *Acacia* (Leguminosae, Mimosoideae). *Ann. Bot.* 50: 721–727.
Kluge, A. G. 1985. Ontogeny and phylogenetic systematics. *Cladistics* 1: 13–27.
Knox, R.B. 1984. The pollen grain. In B.M. Johri, ed. *Embryology of Angiosperms* pp. 197–271. Berlin: Springer-Verlag.
Knox, R.B. and J. Kenrick, 1983. Polyad function in relation to the breeding system of *Acacia.* In D. Mulcahy and E. Ottaviano (eds), *Pollen Biology and Implications for Plant Breeding* pp. 411–417. Amsterdam: Elsevier.
Koponen, T. 1968. Generic revision of the Mniaceae Mitt. (Bryophyta). *Ann. Bot. Fenn.* 5: 117–1551.
Kress, W.J. 1986. Exineless pollen structure and pollination systems of tropical *Heliconia* (Heliconiaceae). In S. Blackmore and I.K. Ferguson (eds), *Pollen and Spores: Form and Function.* London: Academic Press.
Kress, W.J. and D.E. Stone, 1982. Nature of the sporoderm in monocotyledons, with special reference to the pollen grains of *Canna* and *Heliconia. Grana* 21: 129–148.
Kress, W.J. and D.E. Stone, 1983. Morphology and phylogenetic significance of exineless pollen of *Heliconia* (Heliconiaceae). *Syst. Bot.* 8: 149–167.
Kubitzki, K. 1981. The tubular exine of Lauraceae and Hernandiaceae, a novel type of exine structure in seed plants. *Pl. Syst. Evol.* 1381: 139–146.
Kuprianova, L.A. 1979. On the possibility of the development of tricolpate pollen from monosulcate. *Grana* 15: 117–125.
Larson, D.A. 1966. On the significance of the detailed structure of *Passiflora caerulea* exines. *Bot. Gaz.* 127: 40–48.
Longly, B. and L. Waterkeyn, 1979. Étude de la cytocinese. III. Les cloisonnements simultanes et successifs des microsporocytes. *La Cellule* 73: 65–80.
Linder, H.P. and I.K. Ferguson, 1985. Notes on the pollen morphology and phylogeny of the Restionales and Poales. *Grana* 24: 65–76.
Lugardon, B. 1976. Sur la structure fine de l'exospore dans les divers groupes de Ptéridophytes actuelles (microspores et isospores). In I.K. Ferguson and J. Muller (eds), *The Evolutionary Significance of the Exine.* pp. 231–250. London: Academic Press.
Maheshwari, P. 1950. *An Introduction to the Embryology of Angiosperms.* New York: McGraw-Hill,
Mapes, G. and G.W. Rothwell, 1984. Permineralized ovulate cones of *Lebachia* from the late Paleozoic limestones of Kansas. *Palaeontology* 27: 69–94.
McGlone, M.S. 1978. Pollen structure of the New Zealand members of the Styphelieae (Epacridaceae). *New Zealand J. Bot.* 16: 91–101.
Melville, R. 1981. Surface tension, diffusion and the evolution of pollen aperture patterns. *Pollen Spores* 23: 179–203.

Meyer, N.R. and Yaroshevskaya, A.S. 1976. The phylogenetic significance of the development of pollen grain walls in Liliaceae, Juncaceae and Cyperaceae. In I.K. Ferguson and J. Muller (eds), *The Evolutionary Significance of the Exine.* pp. 91–95. London and New York: Academic Press,

Mishler, B.D. and S.P. Churchill 1984. A cladistic approach to the phylogeny of the "bryophytes." *Brittonia* 36: 406–424.

Muller, J. 1970. Palynological evidence on the early differentiation of angiosperms. *Bot. Rev.* 45: 417–450.

Neidhart, H.V. 1979. Comparative studies of sporogenesis in bryophytes. In G.C.S. Clarke and J.G. Duckett, eds. *Bryophyte Systematics.* pp. 251–280. London: Academic Press.

Nelson, G. 1985. Outgroups and ontogeny. *Cladistics* 1: 29–45.

Nelson, G. and N.I. Platnick, 1981. *Systematics and Biogeography: Cladistics and Vicariance.* New York: Columbia University Press,

Pacini, E., G.G. Franchi, and M. Hesse, 1985. The tapetum: its form, function and possible phylogeny in Embryophyta. *Plant. Syst. Evol.* 149: 155–185.

Pankow, H. 1957. Über den Pollenkitt bei *Galanthus nivalis. Flora* 146: 240–253.

Patterson, G. 1982. Morphological characters and homology. In K.A. Joysey and A.E. Friday, eds. *Problems of Phylogenetic Reconstruction.* pp. 21–74. London and New York: Academic Press.

Pettitt, J.M. 1971. Some ultrastructural aspects of sporoderm formation in Pteridophytes. In G. Erdtman and P. Sorsa, *Pollen and Spore Morphology Plant Taxonomy. Pteridophyta.* Stockholm: Almqvist and Wiksell,

Pettitt, J.M. 1981. Reproduction in seagrasses: pollen development in *Thalassia hemprichii, Halophila stipulacea* and *Thalassodendron ciliatum. Ann. Bot.* 48: 609–622.

Pettitt, J.M. and A.C. Jermy, 1974. The surface coats on spores. *Biol. J. Linn. Soc.* 6:245–257.

Pettitt, J.M., S.C. Ducker, and R.B. Knox, 1978. *Amphibolis* pollen has no exine. *Helobiae Newsletters* 2: 19–22.

Pettitt, J.M., Ducker, S.C. and Knox, R.B. 1981. Submarine pollination in seagrasses. *Science* 244: 135–143.

Praglowski, J. 1976. Schisandraceae Bl. *World Pollen and Spore Flora* 5: 1–32.

Punt, W. and M. Monna-Brands. 1977. The Northwest European Pollen Flora, 8. Solanaceae. *Rev. Palaeobot. Palynol.* 23:1–30.

Raj, B. 1961. Pollen morphological studies in the Acanthaceae. *Grana* 3: 1–108.

Rogers, C.M. and B.D. Harris, 1969. Pollen exine deposition: a clue to its control. *Amer. J. Bot.* 56: 101–106.

Roland, F. 1965. Précisions sur la structure et l'ultrastructure d'une tetrade calymmée. *Pollen Spores* 7:5–8.

Roland, F. 1971. The detailed structure and ultrastructure of an acalymmate tetrad. *Grana* 11: 41–44.

Rowley, J.R. 1964. Formation of the pore in pollen of *Poa annua.* In: H.F.

Linskens, ed. *Pollen Physiology and Fertilization*. pp. 59–69. Amsterdam: North Holland Publishing Company,
Rowley, J.R. 1978. The origin, ontogeny and evolution of the exine. *Proc. IV Int. Palynol. Conf., Lucknow* 1: 126–136.
Rowley, J.R. 1981. Pollen wall characters with emphasis on applicability. *Nord. J. Bot.* 1:357–380.
Rowley, J.R. and D. Southworth, 1967. Deposition of sporopollenin on lamellae of unit membrane dimensions. *Nature* 2133: 703–704.
Rowley, J.R. and J.J. Skvarla, 1970. The pollen wall of *Canna* and its similarity to the germinal aperture of other pollen. *Amer. J. Bot.* 57: 519–529.
Rowley, J.R. and J.J. Skvarla, 1975. The glycocalyx initiation of exine spinules on microspores of *Canna. Am. J. Bot.* 62: 479–485.
Rowley, J.R. and A.O. Dahl, 1977. Pollen development in *Artemisia vulgaris* with special reference to glycocalyx material. *Pollen Spores* 19: 169–284.
Sampson, F.B. 1977. Pollen tetrads of *Hedycarya arborea* (Monimiaceae). *Grana* 16: 61–73.
Schopf, J.M. 1938. Spores from the Herrin (No. 6) coal bed in Illinois. *Rep. Invest. Illinois State Geol. Surv.* 50: 1–55.
Schnaar, R.L. 1985. The membrane is the message. *The Sciences* 25: 34–40.
Scott, A. and W.G. Chaloner, 1983. The earliest fossil conifer from the Westphalian B of Yorkshire. *Proc. Roy. Soc. Ser. B.* 220: 13–182.
Shaw, G. 1971. The chemistry of sporopolllenin. In J. Brooks, P.R. Grant, M. Muir, P. van Gijzel and G. Shaw, eds. *Sporopollenin*. pp 305–348. London and New York: Academic Press.
Skvarla, J.J. and D.A. Larson, 1963. Nature of cohesion within pollen tetrads of *Typha latifolia. Science* 140: 173–175.
Skvarla, J.J. and D.A. Larson, 1965. An electron microscopic study in the Compositae with special reference to the Ambrosiinae. *Grana* 6: 210–269.
Skvarla, J.J. and D.A. Larson, 1966. Fine structural studies of *Zea mays* pollen. 1. Cell membranes and exine ontogeny. *Am. J. Bot.* 53: 1112–11125.
Skvarla, J.J. and B.L. Turner, 1966. Systematic implications from electron microscopic studies of Compositae pollen—a review. *Ann. Missouri Bot. Gard.* 5: 220–256.
Skvarla, J.J., P.H. Raven, and J. Praglowski, 1976. Ultrastructural survey of Onagraceae pollen. In I.K. Ferguson and J. Muller (eds.), *The evolutionary Significance of the Exine* pp. 447–480. London: Academic Press.
Skvarla, J.J., B.L. Turner, V.C. Patel, and A.S. Tomb, 1978. Pollen morphology in the Compositae and in morphologically related familes. In V.H. Heywood, J.B. Harborne, and B.L. Turner, eds. *The Biology and Chemistry of the Compositae,* pp. 147–217. London and New York: Academic Press,
Smith White, S. 1959. Pollen development patterns in the Epacridaceae. *Proc. Linn. Soc. N.S.W.* 84: 8–35.
Smith, M.M. and M.E. McCully, 1978. A critical evaluation of the specificity of aniline blue induced fluorescence. *Protoplasma* 95: 229–254.

Southworth, D. 1973. Cytochemical reactivity of pollen walls. *J. Histochem. Cytochem.* 21: 73–80.
Stevens, P. 1980. Evolutionary polarity of character states. *Ann. Rev. Eco. Syst.* 11: 333–358.
Stone, D.E. 1985. Wall ontogeny of Zingiberales and Laurales pollen. *Third International Congress of Systematic and Evolutionary Biology, Abstracts.* p. 185.
Stone, D.E., S.C. Sellers, and W.J. Kress, 1979. Ontogeny of exineless pollen in *Heliconia,* a banana relative. *Ann. Missouri Bot. Gard.* 66:701–730.
Stone, D.E., S.C. Sellers, and W.J. Kress, 1981. Ontogenetic and evolutionary implications of a neotonous exine in *Tapeinochilos* (Zingiberales: Costaceae) pollen. *Amer. J. Bot.* 68: 49–63.
Strandhede, S.O. 1973. Pollen development in the *Eleocharis palustris* subgroup (Cyperaceae). II. Cytokinesis and microspore degeneration. *Bot. Not.* 126: 255–265.
Takahashi, H. and K. Sohma, 1984. Development of pollen tetrads in *Typha latifolia* L. *Pollen Spores:* 26: 5–18.
Taylor, T.N. and G.W. Rothwell, 1982. Studies of seed fern pollen: development of the exine in *Monoletes* (Medullosales). *Am. J. Bot.* 69: 570–578.
Taylor, T.N. and K.L. Alvin, 1984. Ultrastructure and development of Mesozoic pollen: *Classopolis. Am. J. Bot.* 71: 575–587.
Tran Thanh van, K., P. Toubart, A, Cousson, A.G. Darvill, D.S. Gollin, P. Chelf, and P. Albersheim 1985. Manipulation of the morphogenetic pathways of tobacco explants by obligosaccharins. *Nature* 314: 615–617.
Van Campo, M. and Guinet, P. 1961. Les pollens composées. L'example des Mimosacées. *Pollen et Spores* 3: 201–218.
Voorzanger, B. and W.J. van der Steen, 1982. New perspectives on the biogenetic law. *Syst. Zool* 31: 202–205.
Walker, J.W. 1974. Aperture evolution in pollen of primitive angiosperms. *Am. J. Bot.* 61: 891–902.
Walker, J.W. 1976. Evolutionary significance of the exine in the pollen of primitive angiosperms. In I.K. Ferguson and J. Muller ([eds), *The Evolutionary Significance of the Exine,* pp. 251–308. London: Academic Press.
Walker, J.W. and J.A. Doyle, 1975. The bases of angiosperm phylogeny: palynology. *Ann. Mo. Bot. Gard.* 62: 664–723.
Warner, B.G. 1984. Morphological comparison and taxonomic implications of pollen of the Ericales of Canada. *VI International Palynological Conference, Calgary* 1984. Abstract.
Waterkeyn, L. 1962. Les parois microsporocytaires de nature callosique chez *Helleborus* et *Tradescantia. La Cellule* 62: 225–255.
Waterkeyn, L. and A. Bienfait, 1970. On a possible function of the callosic special cell wall in *Ipomoea purpurea* (L.) Roth. *Grana* 10: 13–20.
Watrous, L.E. and Q.D. Wheeler, 1981. The outgroup comparison method of character analysis. *Syst. Zool.* 30: 1–16.

White J. 1979. The plant as a metapopulation. *Ann. Rev. Syst. Ecol.* 10: 109–145.

Wiley, E.O. 1981. *Phylogenetics: the Theory and Practice of Phylogenetic Systematics*. New York: Wiley-Interscience.

Wodehouse, R.P. 1935. *Pollen Grains*. New York: McGraw-Hill.

Zavada, M.S. 1983a. Pollen wall development of *Zamia floridiana*. *Pollen Spores* 25: 287–304.

Zavada, M.S 1983b. Comparative morphology of monocot pollen and evolutionary trends of apertures and wall structures. *Bot. Rev.* 49,: 331–378.

Zavada, M.S. 1984. Pollen wall development of *Austrobaileya maculata*. *Bot. Gaz.* 145: 11–21.

5. Relationships Between Ontogeny and Phylogeny, with Reference to Bryophytes

Brent D. Mishler

Development has become an increasingly central focus for evolutionary biology. Investigations are proceeding into the genetic basis for evolutionary novelties (Raff and Kaufman 1983), ontogenetic connections to life history studies (Stearns 1982) and phenotypic plasticity (Smith-Gill 1983), patterns and mechanisms of heterochronic change (Alberch et al. 1979; Ambros and Horvitz 1984), and developmental canalization as a constraint on natural selection (Alberch 1980, 1982). Systematists currently have a revived interest in relating ontogeny to methods of assessing phylogenetic relationships, in particular for hypothesizing homology, establishing transformation series, and determining evolutionary polarities.

Early references to these important systematic considerations can be found in Gould (1977) and Kluge and Strauss (1985). I am concerned here with assessing the applicability of recent formulations of the relationship between ontogeny and systematics (Nelson 1978, 1985; Fink 1982; Alberch 1985; Brooks and Wiley 1985; De Queiroz 1985; Kluge 1985; Kluge and Strauss 1985; O'Grady 1985) to plants in general and bryophytes in particular. My goals are to: (1) discuss special features of plant development that may either cause difficulties or provide new insights into these considerations, (2) illustrate the apparent importance of ontogenetic modifications in bryophyte phylogeny, and (3) examine the utility of the ontogeny criterion for determining evolutionary polarity.

SPECIAL CONSIDERATONS IN RELATING ONTOGENY TO PHYLOGENY IN PLANT

General differences between growth and development in complex multicellular plants and animals can be stated, although exceptions (and their implications) should always be kept in mind. Five major phenomena are worthy of mention, since they present some difficulties to recent theoretical discussions based primarily on animals. These are: modular growth, hierarchical development of modules, indeterminate growth, phenotypic plasticity, and regeneration.

Most multicellular plants are modular organisms (White 1979). Rather than the complex morphogenetic cell movements and transformations characteristic of unitary animal development, cell division and differentiation occur in repeated units (Tomlinson 1984). Within each unit, cell division and differentiation usually proceed in a comparatively simple gradient from a restricted growing point (the meristem). Modular growth is one factor that has encouraged botanical interest in transformational homology (Stevens 1984), and makes difficult or impossible the application of the "conjunction test" for determining homology. Such a test rejects, as a homology between organisms, characters that co-occur in a single organism (Patterson 1982). This test clearly is inappropriate for plants, in which virtually all possible characters occur repeatedly within individuals. Botanists interested in taxic homology need to use just two tests: similarity and congruence (Patterson 1982).

Most multicellular plants are not modular on only one level. Rather, the modular growth is organized hierarchically within individuals. For example, in moss gametophytes, individual cells (the lowest-level module) have their own pattern of differentiation; each leaf develops according to a certain regular plan; the sequence of leaves produced along a single developing shoot or branch proceeds in a specific and repeatable pattern (Mishler 1986); and even a regular developmental progression in types of shoots produced by a single protonemal system has been noted (Meusel 1935). Thus in this case development is occurring in at least four nested levels of modules. One must therefore be very careful in describing and comparing development to keep track of proper levels.

Indeterminate growth is widely recognized as a general difference between complex multicellular plants and animals (Stevens 1980). Development in plants does not usually cease at the attainment of some specific size or maturity. However, it is important to keep track of modules at different levels here as well, because often certain modules are determinate in their growth while others are indeterminate. Production of floral structures will often terminate the growth of a particular module; further elaboration of the plant then occurs through growth of equivalent modules (Hallé, Oldeman, and Tomlinson 1978). In such a case, the first-order module (a branch) is determinate, the second-order module (a tree) is indeterminate. Since models that have been proposed to study changes in developmental timing during evolution (heterochrony) require fixed and comparable start- and end-points (Alberch et al. 1979; Fink 1982), these models are not appropriate at least for indeterminate growth in plants. However, the applicability of these models to plant organs that are determinate in their growth has been shown by Guerrant (1982).

A further property of plants that causes difficulty in extrapolating to plants from models developed in zoology is their intrinsic phenotypic plasticity. One essential feature of the heterochronic processes modeled by Alberch et al. (1979) and Fink (1982) is the existence of some specific and repeatable rate for the development of a feature. Developmental rates of most modules in plants vary greatly in different environments, making it difficult (or even meaningless) to establish the "true" rate. My own studies of plasticity in mosses (Mishler 1985a) have convinced me that rates of development at all levels can vary extraordinarily under simple and natural variation in light, temperature, and water relationships. Probably the best that can be hoped for is to compare development between species grown in a series of different environments. The developmental "norms of reaction" (Gupta and Lewontin 1982) thus discovered can then be compared to check for consistent similarities and differences between species across environments.

A final topic of perhaps especial interest in plants is regeneration. While dedifferentiation occurs in animals (Kluge 1985), it is ubiquitous enough in most plant groups to be called a way of life. For example, in the bryophytes, virtually every organ can be experimentally induced to dedifferentiate and regenerate new organs of the same or different types. Regeneration is often seen in natural conditions and appears to be very important in asexual reproduction. Giles (1971) has reviewed the process of dedifferentiation in bryophytes and showed that actual reversal of development occurs at both the cellular and organelle levels. Such developmental reversals clearly meet the conditions for falsification of the biogenetic law set by Nelson (1978; cf. Kluge 1985).

The study of ontogeny in relation to phylogeny in plants is thus both simpler in some senses and more complicated in other senses than in animals. The relatively simple and straightforward nature of plant development makes it rather easy to characterize, at least at the level of phenotypic structures (establishing underlying genetic and epigenetic programs remains poorly known for the most part). On the other hand, the same simplicity may make it harder to construct robust hypotheses of phylogeny for plant groups. The modular and hierarchical nature of plant development can result in extensive "mosaic evolution." Given that one can establish reliable cladograms for at least some plant groups, however, the special features of their development can provide new tests for existing models relating ontogeny and phylogeny.

ONTOGENY AND PHYLOGENY IN BRYOPHYTES

The bryophytes, while certainly not a monophyletic group (Mishler and Churchill 1984, 1985), can be usefully grouped together informally for discussion because of a number of shared (plesiomorphic!) structural and ecological factors. This diverse group (containing some 25,000 species) of several early-diverging lineages of land plants is generally characterized by small size and a life-cycle of alternating free-living haploid gametophytes and sporophytes that are epiphytic and partially parasitic on the gametophytes. Generally there is a greater reliance on external water films for sexual reproduction and water conduction than in other land plants such as the angiosperms. Asexual reproduction by regeneration appears to be a major factor in establishment in nature (Mishler in press).

As pointed out by Fink (1982), there is little to be gained in the evolutionary study of ontogeny without having a cladistic hypothesis for the taxa involved. The assumed phylogenetic pattern has everything to do with the type of evolutionary explanations that are given. Such study of ontogeny is hampered in the bryophytes, as in other plants, by the scarcity of cladistic studies (exceptions being Koponen 1968; Bremer 1981; Churchill 1981, 1985; Mishler and Churchill 1984, 1985; Mishler 1985b). In the present discussion I will concentrate on some examples of gametophyte development in bryophytes, drawn from my own and others' observations. I will compare developmental sequences to cladistic hypotheses (where possible), but my major goal is simply to demonstrate the potential for such comparisons in bryophytes, not to draw any firm conclusions.

Developmental changes during the development of leafy shoots

As mentioned above, it is possible to study development at a number of different levels in bryophytes, and parallels between ontogeny and phylogeny might be expected a priori at any of these levels, due to the evident conservatism of development. Areas that have been of particular interest in gametophyte vegetative development in bryophytes include: patterns of germination of spores, protonemal types, and early development of thalli or leafy shoots (termed "gametophores" to distinguish them from the protonemal phase—the gametophyte is thus made up of two major developmental stages); branching patterns of gametophores; and development of individual leaves (e.g., Crandall-Stotler 1981; Frey 1970; Nishida 1978). A relatively neglected area is the sequence of leaves produced along a stem as it matures (Watson 1971). Most leafy bryophytes produce a prolonged sequence of juvenile leaves before mature leaves characteristic of the species are developed. Given certain peculiarities about the way mosses grow, I have argued that such sequences may be of particular interest from a phylogenetic standpoint (Mishler 1986). I will begin with a rather detailed description of this phenomenon in the moss genus *Tortula,* then present a number of other examples in somewhat less detail.

Tortula. A detailed study of ontogeny of leafy shoots relative to cladistic relationships within the genus has recently been published (Mishler 1986), so I will just summarize the major points here. A schematic representation of one species is shown in figure 5.1. The first leaves produced on a shoot are of a form completely unlike that of mature leaves found anywhere in the genus (but, interestingly, rather like the form of mature leaves in other families, such as the Funariaceae or Splachnaceae). These leaves are relatively broad, obtuse, and have plane margins (fig. 5.1a). The leaf cells are large, roughly rectangular or hexagonal, and have smooth surfaces. The costa (midrib) is only differentiated in the lower half of the leaf. Leaves produced later on the stem (fig. 5.1b) are relatively longer, acute, and apiculate, but still with plane margins. Two types of

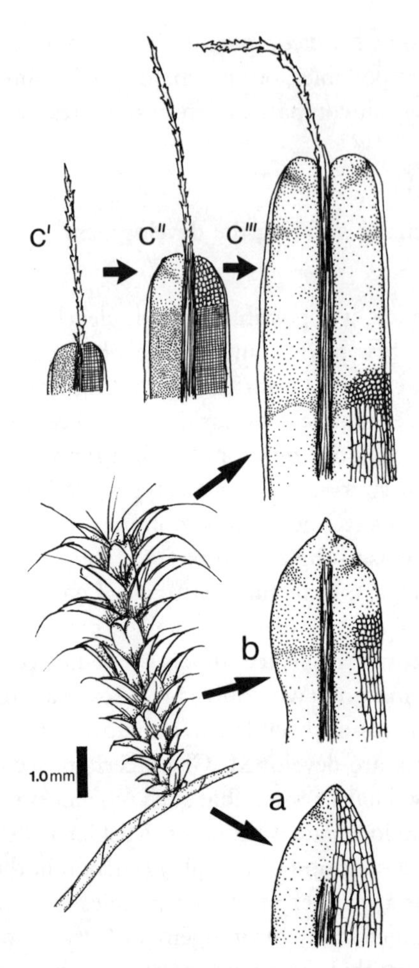

Figure 5.1. Schematic representation of gametophore ontogeny in *Tortula obtusissima* based on culture studies described by Mishler (1986): *a*, *b*, and *c'''* are fully developed leaves of different developmental stages; *c'*-*c'''* shows the development of an individual mature leaf. Reprinted by permission of the editors of *Systematic Botany* 11: 198, 1986.

leaf cells are present: large, rectangular, smooth cells (as in the earlier leaves) in the lower half of the leaf, grading into small, isodiametric and papillose cells above. The costa is differentiated nearly to the leaf apex. The final leaf-form produced in this species (fig. 5.1c''') is much longer in relation to its width, the apex is obtuse, and the leaf margins are recurved nearly to the apex. The two types of leaf cells are strongly differentiated from each other, with virtually no transition zone. The costa is well developed and projects beyond the apex of the leaf as a serrate, hyaline (clear) hairpoint that may be longer than the leaf blade itself.

This developmental sequence, that of the shoot, can be contrasted with that of the lower-order module, the individual leaves. The latter sequence is shown in figure 5.1 (c′-c‴) and is remarkably different from the former sequence. Differentiation in embryonic leaves proceeds from the top down. Young leaves (fig. 5.1c′) have a fully developed hairpoint before the remainder of the leaf is differentiated. The leaf margins are already recurved, however. The next part of the leaf to differentiate is the small upper cells (fig. 5.1c″). Finally the large basal cells are differentiated (fig. 5.1c‴). Every leaf on the shoot (e.g., fig. 5.1a and b) goes through the same "top to bottom" differentiation, although obviously the appearance of features depends on the developmental stage of the stem itself.

These different development patterns at different hierarchical levels within one plant can lead to confusion in description. Standardized terms are needed. I will define "juvenile leaves" as those produced on immature shoots; "mature leaves" to be the final leaf-form produced on normal shoots of a particular species. Early and late developmental stages of individual leaves will be called "young" and "old." Thus one might speak of young or old juvenile and mature leaves.

One additional observation made by Mishler (1986) that bears on the evolution of development in mosses is that the ontogenetic sequence described above for development of shoots from protonema also occurs in the normal development of new branches from mature shoots. In other words, an individual repeats its ontogeny every time it branches. This seems to be a general, if not universal, feature of mosses; thus observations for a number of the examples given below were made from branches on herbarium specimens.

In Mishler (1986), I presented developmental studies of 11 species of *Tortula,* and noted many similarities between juvenile leaves of *T. obtusissima* and mature leaves of other species. Upon comparing these similarities to a cladogram of the genus (which was based on out-group comparison; Mishler 1985b), a number of these similarities appear to be due to recapitulation, others to paedomorphosis.

An example of recapitulation is the likelihood (based on out-group comparison) that gradually differentiated basal cells are plesiomorphic relative to the sharply differentiated basal cells of *T. obtusissima* and a few closely related species. Many of the characteristics of the first juvenile leaves may be recapitulations as well. The form of these leaves is very similar throughout *Tortula* and indeed similar to that of other mosses (as will be mentioned below). These early juvenile leaves have leaf-cells that are reminiscent of leaf cells characteristic of other moss families, notably Bryaceae and Funariaceae. It is of interest that it has been suggested (e.g., Crosby 1980) that these families are close to the "ancestral form" of the subclass Bryidae. However, the significance of these similarities can be assessed only when a detailed cladistic hypothesis for the mosses is available.

A number of instances of paedomorphosis were detected by Mishler (1986). Several species (as judged by out-group comparison and character congruence) appear to have primitively possessed a hairpoint and then lost it, thus resembling figure 5.1b in this respect. One example is *T. fragilis,* a species that is widespread in Mexico but that seems to primarily reproduce asexually because sporophytes are lacking throughout most of its range. It was suggested (Mishler 1986) that the apparent neotenic evolution of *T. fragilis* was duc to selection for rapid asexual reproduction by dedifferentiation and regeneration of leaf cells. It can be experimentally demonstrated that juvenile leaves of species like *T. obtusissima* (and mature leaves of *T. fragilis!)* are much more efficient at this than mature leaves of these species. In any event, a number of other cases of paedomorphosis were discovered (Mishler 1986) involving characters such as the recurvature of the leaf margin, length of the costa, and papillae on the leaf cells.

This particular hierarchical level of development (that of the sequence of leaves produced along a shoot) appears to be particularly relevant to phylogenetic comparisons among species because many features of this sequence appear in mature leaves of related species. This is in contrast to the development of individual leaves (fig. 5.1, c′–c‴) where such parallels were not discovered. The prolonged ontogeny of leafy shoots, involving characters of phylogenetic importance, and the frequency of apparent heterochronic events in the evolution of *Tortula,* suggest that this level of development may be of particular interest in studies of other mosses.

Fissidens. The bizarre leaves occurring in this genus have long been a source of interest and phylogenetic speculation. The mature leaves have three laminae, instead of the normal two found in other mosses. The lower part of the leaf appears "Y" shaped in cross section, with two laminae (the vaginant laminae) that clasp the stem and another lamina that projects on the other side of the costa from these (see the late juvenile leaf in fig. 5.2f). The region at the top of the leaf beyond where the vaginant laminae join is sometimes called the apical lamina. While undoubtedly a uniquely derived feature, the transformational homologies of this leaf to normal moss leaves is far from obvious.

Robinson (1970) has reviewed the history of interpretations of the *Fissidens* leaf-type. Some have interpreted the dorsal + apical + one vaginant laminae as the "true leaf" and viewed the other vaginant lamina as a duplication or a fusion of an additional leaf. Others (most clearly shown by Salmon 1899) have considered the vaginant laminae to represent the "true leaf" and the dorsal and apical laminae to represent outgrowths from the ancestral simple leaf.

An analysis of the sequence of juvenile leaves produced on branches of *Fissidens* (mentioned, among others, by Salmon 1899) gives convincing support for the latter interpretation (fig. 5.2). Early juvenile leaves (fig. 5.2a) are like normal moss leaves in every way, and resemble leaves of families thought to be closely related because of sporophyte characters (such as Ditrichaceae and Dicranaceae). The leaves clasp the stem, are obtuse, and have an apiculate point. Later leaves (fig. 5.2b) are acute and have a more prominent point. Leaves produced just after this stage (fig. 5.2c) have an obviously fleshy point with larger cells than the laminae. Successive juvenile leaves (fig. 5.2d-f) have recognizable apical and dorsal laminae, both becoming more elaborate in later leaves.

This sequence clearly supports the hypothesis that the dorsal and apical laminae were derived as outgrowths from the apex and costa of the normal moss leaf-type. A further likely recapitulation involves the very rare type of apical cell characteristic of *Fissidens,* a two-sided lenticular cell which leads to a distichous leaf arrangement (Chamberlin 1980). This type of apical cell is certainly a derived character for the genus, based on any likely out-group comparison, as compared to the common tetrahedral apical cell possessed by most mosses. A parallel transition occurs ontogenetically as well, since young shoots of *Fissidens* possess a tetrahedral apical cell which transforms to the lenticular one (Chamberlin 1980).

Atrichum. This genus is a member of the Polytrichales, an order that

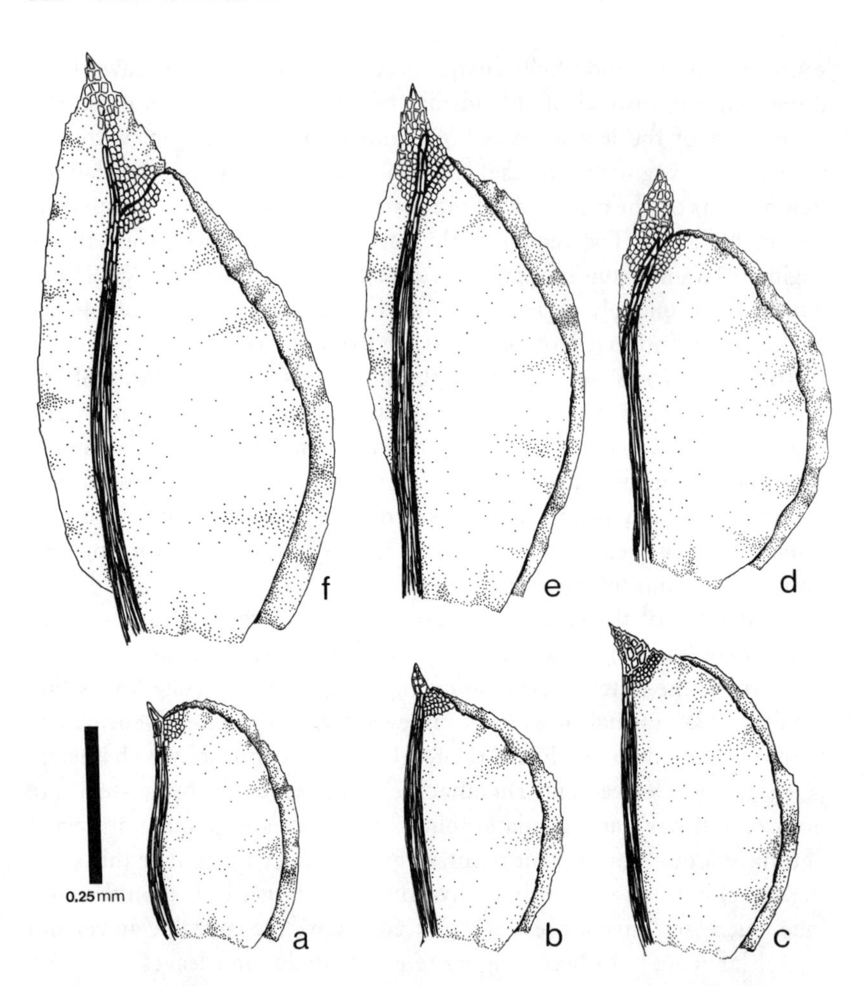

Figure 5.2. A selection of fully developed juvenile leaves of *Fissidens cristatus* taken from successive positions along a developing branch: *a* occupied the lowest (earliest) position on the branch, *f* the highest (latest). Note that *f* is not yet the mature leaf-form of this species; these would have a relatively longer apical lamina and a larger dorsal lamina [voucher: *Mishler* 3730, North Carolina, 1 Nov. 1985, DUKE].

appears to be relatively plesiomorphic compared to the majority of mosses (Mishler and Churchill 1984). However, the order appears closer cladistically to the true mosses (represented above by *Fissidens* and *Tortula)* than to *Sphagnum* (discussed below). Leaves of most of the Polytrichales have lamellae on the surface of the costa (figure 5.3A, c). These are photosynthetic outgrowths of the costa that, on the basis of the above discussion, are similar in some ways (but not homologous) to the dorsal lamina of *Fissidens*. Lamellae are known elsewhere in the mosses, but are not found in close outgroups to the Polytrichales; therefore they are likely to be derived for the order.

Examination of the development of leafy shoots of *Atrichum* shows apparent recapitulations of several evolutionary transformations (fig. 5.3A). The early juvenile leaves (fig. 5.3A, a) are like those found widely in mosses (see fig. 5.1a), with short costa and large, smooth leaf-cells. Later juvenile leaves (fig. 5.3A, b) develop the border of elongated cells. This is probably synapomorphic for the genus since it is not found in other genera of the family (Smith 1971) or in close outgroups. Still later juvenile leaves (fig. 5.3A, c) develop the characteristic photosynthetic lamellae. Note that there is some apparent incongruence here between ontogeny and phylogeny, in that the leaf border (synapomorphic for the genus) appears well before the costal lamellae (synapomorphic for the order).

Sphagnum. The leaves of *Sphagnum* are as bizarre in their own way as those of *Fissidens,* but again it appears that examination of the development of leafy shoots can clarify transformational homologies (fig. 5.3B; also illustrated nicely by Schimper, 1858). The mature leaves of all species of *Sphagnum* have leaf cells that are strongly differentiated into two types (fig. 5.3B, c): enlarged, empty cells usually with pores and reinforced with annular fibrils (responsible for the legendary water-holding capacity of *Sphagnum),* surrounded by smaller, narrow, living cells packed with chloroplasts.

The earliest juvenile leaves (fig. 5.3B, a) however, have cells of only one type: relatively large and rectangular cells, all of which contain chloroplasts. Subsequent juvenile leaves (fig. 5.3B, b) very soon have a noticeable distinction between larger clear cells, devoid of chloroplasts but without annular fibrils and somewhat narrower chlorophyllous cells. Later juvenile leaves (fig. 5.3B, c) develop the border of linear cells which are characteristic of a number of species of *Sphagnum.*

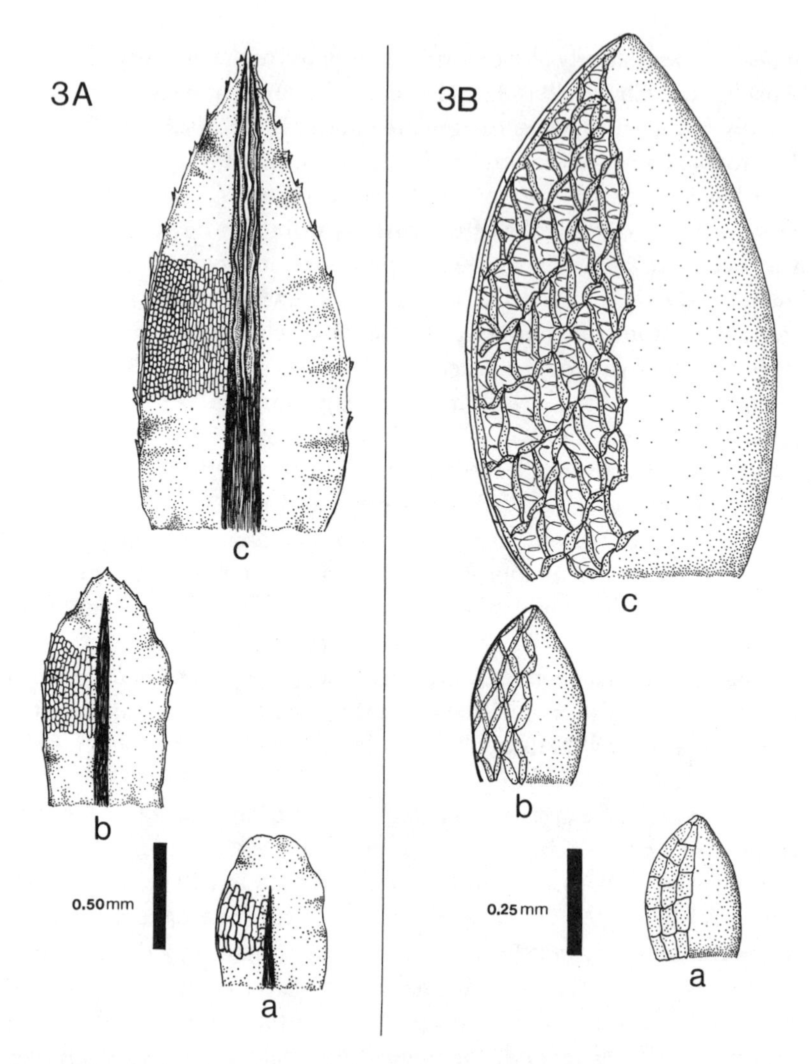

Figure 5.3. 3A. A series of juvenile leaves from a developing branch in *Atrichum angustatum*. *a* occupied the lowest (earliest) position on the branch, *c* the highest (latest) [voucher: *Mishler* 3743, North Carolina, April 1985, DUKE]. 3B. A series of juvenile leaves from a developing shoot (grown in culture) in *Sphagnum compactum*: *a* occupied the earliest position on the shoot, *c* the latest [voucher: Risk 751, North Carolina, March 1986, DUKE].

Since the unique leaf cell dimorphism of *Sphagnum* is certainly a synapomorphy, this ontogenetic sequence appears to represent an instance of recapitulation. Another nice example of recapitulation is provided by the presence of normal moss rhizoids in juvenile plants of *Sphagnum* (an important synapomorphy of mosses), that are lacking in mature plants (Mishler and Churchill 1984).

Other examples. Many additional examples of apparent relationships between ontogeny and phylogeny in the development of bryophyte gametophores could be cited, and indeed it is beginning to look as though careful developmental study of most groups will yield valuable comparative information relevant to all phylogenetic levels. A few more examples could be instructive.

Churchill (1985) presented a cladistic analysis of the three species of the moss genus *Scouleria*. He postulated that the multistratose leaf border (pseudocosta) shared by two species, *S. marginata* and *S. patagonica* is a synapomorphy. This polarity decision is well-supported by outgroup comparison, but it is interesting to note that it is well-supported by the ontogeny criterion as well. Examination of developing branches in herbarium specimens [*S. marginata,* Idaho, Young 292, DUKE. *S. patagonica,* Chile, Crosby 12392, DUKE.] shows that juvenile leaves in both species lack the pseudocosta entirely.

The moss *Eustichia* has an unusual distichous habit, with two opposing rows of large, clasping leaves along the stem. A third row of rudimentary leaves is present in mature stems. Examination of the leaf arrangement at the base of new branches [*E. spruceana,* Guatemala, Steyermark 35703, DUKE.] demonstrates, however, that the juvenile leaves are arranged spirally around the stem and all three rows of leaves are equal in size and shape, which is quite likely to be a recapitulation of the ancestral condition. The juvenile leaves also differ significantly from the mature leaves in shape and cellular details; study of the juvenile leaves therefore might be decisive in determining the relationships of the genus (which remain controversial).

Renzaglia and Bartholomew (1985) have recently studied a similar transformation in the hepatic genus *Fossombronia,* which is characterized by two-ranked "leaves" and a lenticular apical cell. However, it is likely based on both developmental comparisons and phylogenetic congruence that these leaves are homologous neither to those of the mosses or the true leafy hepatics, the Jungermanniales (Renzaglia 1982; Mishler and

Churchill 1984, 1985). Renzaglia and Bartholomew have determined that in early stages of gametophore development a tetrahedral apical cell is found, and three ranks of leaves are initially formed. The transition to a lenticular apical cell and two-ranked leaves can be directly observed, and they correctly pointed out that this observation implies a similar phylogenetic transformation. Such a phylogenetic polarity has support from outgroup comparison as well, since it appears that radical symmetry seems to be primitive for the Metzgeriales (to which *Fossombronia* belongs—see cladistic analysis by Mishler and Churchill 1985).

One final example involves symmetry of leaves themselves. The moss family Neckeraceae is generally characterized by asymmetric leaves, i.e. with unequal laminae. However, as pointed out to me by Sastre-De-Jesus (pers. comm.), the juvenile leaves of both *Neckera* and *Neckeropsis* are symmetric and differ in a number of other respects from the mature leaves. In this as in other families of pleurocarpous mosses where the length and other features of the costa are important taxonomic characters, the examination and comparison of juvenile leaves may be especially significant since extensive variation in costae is often seen.

Protonemal systems

The elaborate filamentous system which is produced as the first stage of the gametophyte in mosses has sometimes been viewed as a grand recapitulation itself—a remnant of the green algal ancestry of the land plants. However, this is one of those stories that, while so good they deserve to be true, are not true when examined closely. Out-group comparison, character congruence, and parsimony strongly support the view that the elaborate protonemal system of the true mosses is a synapomorphy as compared to the plesiomorphic short-filamentous or thalloid protonema of all other land plants and closely related groups of green algae (Mishler and Churchill 1984, 1985). Thus we have here a clear case of nonterminal addition to a developmental sequence (O'Grady 1985).

The basic moss protonema has been evolutionarily modified in various lineages within the mosses (for a detailed survey of the range of variation see Nishida 1978). Modifications include reduction, endosporic germination, production of thalloid structures before true gametophores, and "tubers." The latter organs are clusters of cells borne on rhizoids and protonemata that appear to function in asexual reproduction. Each type of specialization is of course nonterminal, and involves either deletion, addition, or substitution in the sense of O'Grady (1985), and provides further falsification of the biogenetic law.

To summarize, gametophore development of both mosses and liverworts exhibits orderly sequences of many taxonomically important characters, including nature of the apical cell, shape and arrangement of leaves, types of leaf cells, costae, leaf border, and many others. This is a largely untapped source of taxonomic characters, evidence for transformational homology, and potential use for polarizing transformation series. However, the literal application of "ontogeny recapitulates phylogeny" should be avoided since paedomorphosis and nonterminal ontogenetic changes have been demonstrated.

DISCUSSION AND CONCLUSION

Nelson (1978) formulated the biogenetic law in a way that was directly falsifiable (see Weston, present volume). A number of recent papers have documented the flaws of Nelson's formulation, at least as a universal law (Alberch 1985; Brooks and Wiley 1985; De Queiroz 1985; Kluge 1985; Kluge and Strauss 1985; O'Grady 1985), so there is no need for detailed discussion here. Additional cases of falsification of the biogenetic law (at least if the law is interpreted in such a way as to be falsifiable—cf. Kluge 1985) presented here include: (1) dedifferentiation and regeneration, a ubiquitous process in bryophytes, responsible both for asexual reproduction and normal branching of the plants, that involves directly observable reverses in ontogenetic sequences; (2) the very different phenotypic sequences seen in the differentiation of individual leaves versus the developmental sequence of leaves along a shoot (see under *Tortula,* above—both sequences could not recapitulate phylogeny because they are mutually contradictory; (3) paedomorphic events in the evolution of *Tortula;* (4) nonterminal additions, deletions, and substitutions in the evolution of the moss protonemal phase.

Therefore, I would have to agree with a number of phylogenetic systematists who insist that the fundamental basis for systematics is not ontogeny, as stated by Nelson and others, but rather out-group comparison, congruence, and parsimony. This is not to say, however, that close parallels do not exist between ontogeny and phylogeny in many organisms. The biogenetic "law" may be at least a useful empirical generalization. In groups of organisms such as bryophytes, where cladistic relationships are incompletely known and out-group comparison is often difficult, the ontogeny criterion may be justifiable as a means for initially polarizing characters. Its justification is based on the preliminary observations given above that ontogeny does indeed seem to recapitulate some aspects of phylogeny during shoot development. Further justification in bryophytes or in other organisms will not be possible until a good deal more is understood about developmental processes.

As pointed out by several workers (Roth 1984; Alberch 1985) it will be necessary for systematists to follow the lead of developmental biologists in further study of ontogeny and phylogeny. Our current recognition of discrete phenotypic character-stages in ontogeny, which are homologized to characters of adult organisms of other species, may seem quite naíve when genetic and epigenetic processes underlying phenotypes are known. Kluge and Strauss (1985), De Queiroz (1985), and Creighton and Strauss (1986) have advocated the use of ontogenetic transformations themselves as phylogenetic characters.

For the time being, it would seem that systematists could make use of an ontogenetic perspective in several ways. Early developmental stages can simply be used as a source of new characters, which may be especially useful for groups (e.g., *Sphagnum)* that are so highly autapomorphic in their mature form that few characters remain to show their relationships to other groups. As mentioned above, the ontogeny criterion can be used to tentatively polarize characters when out-group comparisons are not available. Finally, the time-honored use of development as one prime criterion for determining homology is appropriate and useful both for determining taxic and transformational homology. Here too, caution is warranted because nonterminal ontogenetic changes are clearly possible. Thus a feature present in two organisms may have different developmental antecedents in the two organisms yet be homologous in the sense of phylogenetic derivation from a single developmental program in a common ancestor (Roth 1984).

Some problems in relating development and evolution can be approached experimentally, and bryophytes may be especially amenable to such an approach, as shown in the important series of studies by Basile (1969, 1974) and Basile and Basile (1983, 1984). For example, if a paedomorphic event is postulated to have occurred in the phylogenetic history of a species, then it might be possible to experimentally induce the ancestral mature morphology, given a sufficient understanding of regulatory events in development. It has been pointed out that manipulative experiments on development must be interpreted carefully in a phylogenetic context because an "atavism" may instead be a structure produced *de novo* because of the manipulation (O'Grady 1985). Such experiments should of course be based on a thorough knowledge of normal developmental processes and set in an explicit cladistic hypothesis. Nonetheless, the eventual attainment of such knowledge is an exciting goal, and the bryophytes may well be ideal subjects for study.

ACKNOWLEDGMENTS

I thank Mark Chamberlin, Steve Churchill, Karen Renzaglia, Ines Sastre-De-Jesus, and Holly Thacher for contributing examples and discussion. Travel support was provided by the Duke University Research Council.

REFERENCES

Alberch, P. 1980. Ontogenesis and morphological diversification. *Amer. Zool.* 20: 653–667.

Alberch, P. 1982. Developmental constraints in evolutionary processes. Pp. 313–332 In J. T. Bonner, ed. *Evolution and Development*, Berlin: Springer-Verlag.

Alberch, P. 1985. Problems with the interpretation of developmental sequences. *Syst. Zool.* 34: 46–58.

Alberch, P., S. J. Gould, G. F. Oster, and D. B. Wake. 1979. Size and shape in ontogeny and phylogeny. *Paleobiology* 5: 296–317.

Ambros, V. and H. R. Horvitz. 1984. Heterochronic mutants of the nematode *Caenorhabditis elegans. Science* 226: 409–416.

Basile, D. V. 1969. Toward an experimental approach to the systematics and phylogeny of leafy liverworts. Pp. 120–133 *in* J. E. Gunckel, ed. *Current Topics in Plant Science.* New York: Academic Press.

Basile, D. V. 1974. A possible role for hydroxyproline-proteins in leafy liverwort phylogeny. *J. Hattori Bot. Lab.* 38: 91–98.

Basile, D. V. and M. R. Basile. 1983. Desuppression of leaf primordia of *Plagiochila arctica* (Hepaticae) by ethylene antagonist. *Science* 220: 1051–1053.

Basile, D. V. and M. R. Basile. 1984. Probing the evolutionary history of bryophytes experimentally. *J. Hattori Bot. Lab.* 55: 173–185.

Bremer, B. 1981. A taxonomic revision of *Schistidium* (Grimmiaceae, Bryophyta) 3. *Lindbergia* 7: 73–90.

Brooks, D. R. and E. O. Wiley. 1985. Theories and methods in different approaches to phylogenetic systematics. *Cladistics* 1: 1–11.

Chamberlin, M. A. 1980. The morphology and development of the gametophytes of *Fissidens* and *Bryoxiphium* (Bryophyta). M.Sc. Thesis. Southern Illinois University at Carbondale.

Churchill, S. P. 1981. A phylogenetic analysis, classification and synopsis of the genera of the Grimmiaceae (Musci). In V.A. Funk and D. R. Brooks, eds. *Advances in Cladistics,* 1: 127–144. New York Botanical Garden.

Churchill, S. P. 1985. The systematics and biogeography of *Scouleria* Hook. (Musci: Scouleriaceae). *Lindbergia* 11: 59–71.

Crandall-Stotler, B. 1981. Morphology/anatomy of Hepatics and Anthocerotes. *Advances in Bryology* 1: 315–398.

Creighton, G. K. and R. E. Strauss. 1986. Comparative patterns of growth and development in Cricetine rodents and the evolution of ontogeny. *Evolution* 40: 94–106.

Crosby, M. R. 1980. The diversity and relationships of mosses. In R. J. Taylor and A. E. Leviton, eds. *The Mosses of North America,* pp. 115–129. AAAS: Pacific Division.

De Queiroz, K. 1985. The ontogenetic method for determining character polarity and its relevance to phylogenetic systematics. *Syst. Zool.* 34: 280–299.

Fink, W. L. 1982. The conceptual relationship between ontogeny and phylogeny. *Paleobiology* 8: 254–264.

Frey, W. 1970. Blattentwicklung bei Laubmoosen. *Nova Hedwigia* 20: 463–556.

Giles, K. L. 1971. Dedifferentiation and regeneration in bryophytes: a selective review. *N. Z. J. Bot.* 9: 689–694.

Gould, S. J. 1977. *Ontogeny and phylogeny.* Cambridge: Harvard University Press.

Guerrant, E. O. 1982. Neotenic evolution of *Delphinium nudicaule* (Ranunculaceae): a hummingbird-pollinated larkspur. *Evolution* 36: 699–712.

Gupta, A. P. and R. C. Lewontin. 1982. A study of reaction norms in natural populations of *Drosophila pseudoobscura. Evolution* 36: 934–948.

Hallé, F., R. A. A. Oldeman and P. B. Tomlinson. 1978. *Tropical trees and forests: an architectural analysis.* Berlin: Springer-Verlag.

Koponen, T. 1968. Generic revision of Mniaceae Mitt. (Bryophyta). *Ann. Bot. Fenn.* 5: 117–151.

Kluge, A. G. 1985. Ontogeny and phylogenetic systematics. *Cladistics* 1: 13–27.

Kluge, A. G. and R. E. Strauss. 1985. Ontogeny and systematics. *Ann. Rev. Ecol. Syst.* 16: 247–268.

Mishler, B. D. 1985a. Biosystematic studies of the *Tortula ruralis* complex. I. Variation of taxonomic characters in culture. *J. Hattori Bot. Lab.* 58: 225–253.

Mishler, B. D. 1985b. The phylogenetic relationships of *Tortula:* an SEM survey and a preliminary cladistic analysis. *Bryologist* 88: 388–403.

Mishler, B. D. 1986. Ontogeny and phylogeny in *Tortula* (Musci: Pottiaceae). *Syst. Bot.* 11: 189–208.

Mishler, B. D. In press. Reproductive ecology of bryophytes. In J. Lovett Doust and L. Lovett Doust eds. *Reproductive Strategies of Plants.* Oxford University Press.

Mishler, B. D. and S. P. Churchill. 1984. A cladistic approach to the phylogeny of the "bryophytes." *Brittonia* 36: 406–424.

Mishler, B. D. and S. P. Churchill. 1985. Transition to a land flora: phylogenetic relationships of the green algae and bryophytes. *Cladistics* 1: 305–328.

Meusel, H. 1935. Wuchsformen und Wuchstypen der europäischen Laubmoose. *Nova Acta Leopoldina* 3: 123–277.

Nelson, G. 1978. Ontogeny, phylogeny, paleontology, and the biogenetic law. *Syst. Zool.* 27: 324–345.

Nelson, G. 1985. Outgroups and ontogeny. *Cladistics* 1: 29–45.

Nishida, Y. 1978. Studies on the sporling types in mosses. *J. Hattori Bot. Lab.* 44: 371–454.

O'Grady, R. T. 1985. Ontogenetic sequences and the phylogenetics of parasitic flatworm life cycles. *Cladistics* 1: 159–170.

Patterson, C. 1982. Morphological characters and homology. *Syst. Assoc. Special Volume* 21: 21–74. London: Academic Press.

Raff, R. A. and T. C. Kaufman. 1983. *Embryos, genes, and evolution: The developmental-genetic basis of evolutionary change.* New York: MacMillan.

Renzaglia, K. S. 1982. A comparative developmental investigation of the gametophyte generation in the Metzgeriales (Hepatophyta). *Bryophytorum Bibliotheca* 24. Vaduz: J. Cramer.

Renzaglia, K. S. and S. E. Bartholomew. 1985. Sporeling development in *Fossombronia cristula* Aust. with special reference to the apical organization and growth. *Bryologist* 88: 337–343.

Robinson, H. 1970. Observations on the origin of the specialized leaves of *Fissidens* and *Schistostega. Rev. Bryol. et Lichen.* 37: 941–947.

Roth, V. L. 1984. On homology. *Biol. J. Linn. Soc.* 22: 13–29.

Salmon, E. S. 1899. On the genus *Fissidens. Annals of Botany* 13: 103–130.

Schimper, W. P. 1858. *Versuch einer Entwickelungs-Geschichte der Torfmoose.* Stuttgart: E. Schweizerbart.

Smith, G. L. 1971. A conspectus of the genera of Polytrichaceae. *Mem. N.Y. Bot. Garden* 21(3): 1–83.

Smith-Gill, S. J. 1983. Developmental plasticity: developmental conversion versus phenotypic modulation. *Amer. Zool.* 23: 47–55.

Stearns, S. C. 1982. The role of development in the evolution of life histories. In J. T. Bonner, ed. *Evolution and Development.* Berlin: Springer-Verlag. pp. 237–258.

Stevens, P. F. 1980. Evolutionary polarity of character states. *Ann. Rev. Ecol. Syst.* 11: 333–358.

Stevens, P. F. 1984. Homology and phylogeny: morphology and systematics. *Syst. Bot.* 9: 395–409.
Tomlinson, P. B. 1984. Homology: an empirical view. *Syst. Bot.* 9: 374–381.
Watson, E. V. 1971. *The Structure and Life of Bryophytes.* 3d ed. London: Hutchinson.
White, J. 1979. The plant as a metapopulation. *Ann. Rev. Ecol. Syst.* 10: 109–145.

6. Age-Dependent Evolution: From Theory to Practice

Henri M. André

MacBride was in his garden settling pedigrees,
There came a baby Woodlouse and climbed upon his knees,
And said: "Sir, if our six legs have such an ancient air,
Shall we be less ancestral when we've grown our mother's pair?"

Garstang, 1922 (or earlier).[1]

Molting is an important step in the life cycle of arthropods. However, the molting process and the related concepts of instar and stage (stadium) cannot account for the diversity of their life cycles. Indeed, not every molt has the same effect. In many cases, molting results in a form distinctly different from the preceding ones. The process associated with distinct successive forms demands a conceptual framework emphasizing the changes occurring through ontogeny—i.e., a paradigm stressing development rather than growth.

The stase concept answers such a requirement and will be presented in section 1. Section 2 will deal with a paradigm derived from the stase concept, namely the "évolution selon l'âge" theory proposed by Grandjean (see 1957b). This theory, founded on the study of mites, will be discussed and extended to other arthropods. Furthermore, it will be placed in a larger context and associated with the views elaborated by Garstang (1922–28) and those introduced by Hennig (1950–1966). If, as stated by Griffiths (1974), the foundations of systematics lie in ontology, systematics may not ignore the "immature" forms under which many arthropods spend most of their lives, or disregard ontogeny as the dynamic process leading these organisms through successive and discrete steps from birth to reproduction. These views will be discussed in section 3, and two original approaches—namely the ontogenetic dendrogram and the ontogenetic trajectory methods—will be presented to incorporate them into systematics.

The ideas introduced by Garstang and the theory propounded by Grandjean should have altered our whole outlook on evolutionary biology. However, they are still misunderstood. My aim in this essay is to provide biologists and systematists with a concise introduction to these theories and to make a first attempt to put them into practice.

THE BASIC CONCEPTS

We are concerned here with definitions of basic terms and the biological significance associated with them. The instar-stage-stadium terminology is detailed in subsection A, while subsection B describes the stase notion. The last two subsections deal with the basic forms under which a stase may exist.

The instar-stage-stadium terminology

Some confusion has revolved around the different interpretations of an instar and numerous publications have been devoted to standardizing terminology. In spite of some objections (Fink 1983; Jones 1983; Schaefer 1983; Steyskal 1984), a general agreement seems to exist in defining the instar as the arthropod itself or the form assumed by the arthropod between two successive molts (Snodgrass 1935; Imms 1948; Anderson et al. 1971; Carlson 1983). In contrast, the period of life between two successive molts is called the stadium (Imms 1948; Anderson et al. 1971; Jones 1978; Carlson 1983) or stage (Snodgrass 1935; Imms 1948; Anderson et al. 1971). Similarly, the French workers Vachon (1953) and Juberthie (1955) proposed the terms "stase" to replace the English term "instar" and "stade" as an equivalent to the English term "stadium." Fortunately, their proposal has not been followed and the term "stade" is isolated.[2] Furthermore, a general agreement seems also to exist, at least among entomologists, in limiting the definition of the term instar and applying this term only to those instars which occur after leaving the egg, as initially proposed by Carothers (1923).

However, the general agreement disappears once a more precise definition is required. Hinton, in a series of publications (1946a, 1958, 1966a, 1971, 1973, 1976) advocated that the term instar should be related to apolysis rather than ecdysis. Owing to this proposal, problems with the instar, stadium, and stage terminology have developed. Jones (1978) suggested reserving the term instar to designate the animal between two successive apolyses and the use of the term stage when the moment of apolysis is unknown. Fink (1983) proposed the retention and expansion of the original meanings of instar and stadium so that both terms might be used in reference to ecdysis or apolysis. In addition, the ideas propounded by Hinton have been disputed by several authors such as Wigglesworth (1973) and Whitten (1976). In any case, the terms instar, stadium, and stage—whether they refer to apolysis or ecdysis—are clearly defined in reference to the molting processes, and to nothing more.

The core concept of stase

Lubbock (1873) drew a distinction between animals with different terminal or mature forms and animals which pass through a succession of different forms in the course of their development. He proposed that the term polymorphism be restricted to the former phenomenon and polyeidism to the latter (figure 6.1). This distinction is fundamental. As noticed by Wigglesworth (1954), the essential feature of polymorphism is multiple potentiality: one form alone is realized, while the others remain latent or suppressed. Similarly, the essential feature of polyeidism is multiple potentiality but, conversely, different forms coexist in the same animal and succeed one another during the course of its ontogeny. The former involves interindividual variations, the latter to intraindividual ones. A similar view was proposed by Hennig (1950) who did not know of—or at least did not refer to—Lubbock's (1873) book. Indeed, Hennig (1950) designated, under the term allomorphism, the general diversity of forms observed in the same species, and subdivided allomorphism into "Metamorphismus" (equivalent to Lubbock's polyeidism) and "Polymorphismus" (equivalent to Lubbock's polymorphism).[3] It is unfortunate that this essential difference is disregarded by authors who state that polymorphism applies also to the discontinuous morphs exhibited in ontogenetic sequences (Matsuda 1979: 211; Bouhniol 1980: 150).

The successive morphs succeeding one another during development conform with the concept of stase propounded by Grandjean (1938a,

Figure 6.1. Diagrammatic representation of two types of allomorphism: polymorphism (A) vs. polyeidism (B) *sensu* Lubbock (1873).

1951b, 1957b, 1970) who applied it to mites. Although it is also related to molting processes, the stase concept differs in basic ways from those used in the instar-stage-stadium terminology. Indeed, a stase is defined as one of the successive forms through which an arthropod passes, these forms being fundamentally different from one another by the criterion of "all or none." In this definition, two points distinguish the stase from the instar. First, the successive forms are fundamentally different from one another; this difference does not constitute a prerequisite in most definitions of the instar.[4] Second, the nature of the difference is explicitly given in the definition which rests on binary data, i.e., on presence and absence data. An organ, a seta for instance, exists in one stase but is absent from the subsequent one, or the converse.

Another example of relevant characters is provided by oribatid mites where a pair of genital acetabula is added at each nymph stase. Differences in dimensions of existing structures between instars, even if they present discontinuous or nonoverlapping values, are thus rejected as discriminant characters between stases, as are allometric changes. Grandjean (1970) even added that the criterion of "all or none" should apply to idionymic organs.[5] *The change in character is emphasized,* not the change of skin. This clearly means that the stase concept relates to development and not to growth. However, the stase definition does not imply that it is related to metamorphoses or any other drastic changes occurring during development, as wrongly thought by some authors who concluded that the stase concept does not apply very well to certain groups—(e.g., Betsch (1987) for Collembola and Emerit (in lit.) for spiders.

In some cases, the number of stases may differ from the number of instars. These special cases reveal that not every molt has the same effect. Depending on the effect, three kinds of molts have to be distinguished (Grandjean 1970; Hammen 1978, 1980). Repetition molts separate two instars which are identical, even in size. Growing-molts separate two instars presenting allometric changes or any other differences in dimensions. Finally, molts may result in different stases. As a consequence, a stase is always an instar, but not the reverse. Logically, "stase" is a proper subset of the set "instar." Practically, a stase may consist of several instars.

Irrespective of any growing process, the development of an arthropod is thus conceived as a discontinuous process where each step corresponds to a stase. In other words, a stase is *an animal at any level of its ontogeny* (Grandjean 1957b: 483–484; 1970:796). The successive levels can merely be numbered, but usually are designated through a particular terminology involving such terms as prelarva, larva, nymph, adult, etc.[6]

Another fundamental character of the stase, implicit in the original definitions by Grandjean, has been stressed by Grandjean himself (1957a) and Hammen (1975, 1978). Stases are idionymic, which implies that they are identifiable *per se* (it is not necessary to count the number of molts or to make any measurements) and can be homologized with corresponding instars in other species of the same taxon. In Acari, or at least in Actinotrichida, such homologies can be established as the fundamental number of stases is known to be six. For instance, acarologists now know that the so-called hypopus—a special form observed in some acarid mites—is nothing more than a deuteronymph. To use the term deuteronymph to designate a hypopus is tantamount to establishing a homology with any other deuteronymph found in Actinotrichida, however different they are. In other groups, however, such homologies are difficult to determine because the number of successive forms is variable or not yet defined. This is why Hammen (1975) introduced the term "stasoid" to designate forms which differ from one another by distinct discontinuous characters, but which cannot be homologized with corresponding instars in other species of the same group.

The stase concept also presents some similarities with the semaphoront concept advanced by Hennig (1950, 1966) insofar as both deal with diversity of forms. However, they differ both in theory and practice. A semaphoront was explicitly defined in reference to polyeidism by Hennig (1950). The semaphoront is a stage of development, but it is also implicitly related to polymorphism. Indeed, if for instance sexual dimorphism is well-marked in a species, the adult semaphoront of that species, as any individual, must exhibit either male or female characters. As for the stase, this is defined only in reference to polyeidism. In practice, this means that a stase may involve different semaphoronts. In the above example, the adult stase comprises two semaphoronts, the male and female forms. In addition, the existence and duration of a stase are determined by molting processes while, as noted by Hennig (1950), there is no general rule for determining what constitutes a stage or semaphoront. Lastly, irrespective of any polymorphism, a stase may in some rare cases include more than one semaphoront as a result of neosomy, a process defined as the formation of new morphological characters, accompanied by growth of new tissue during a single stase (Audy et al. 1963, 1972; Fain 1967).

The states of a stase

A stase may manifest itself in different ways, that is, under different states. Two special states were emphasized by Grandjean (1938a, 1957b). Elattostasis (1957) describes an animal whose mouth parts are reduced in such a way that it is unable to eat. Classical examples include the phoretic hypopods in mites and by the adult mayflies in insects. Calyptostasis (1938a) differs from elattostasis by describing total loss of appendages or, at least, their use. A calyptostasis is thus a nonfeeding and nonwalking form. The most classical example of such an inhibited state is provided by the chrysalid of butterflies. Very often, a calyptostasis is reduced to a mere skin (cuticle) called an apoderma, a term coined by Henking (1882). The term "pullus" used in entomology and arachnology also describes a special state and is equivalent to both calypto- and elattostasis, or only to calyptostasis[7] depending on the author's usage (see e.g., Canard 1984).

Very often, a calyptostatic stase does not emerge from the skin of the previous one (or, possibly, from the egg envelope). In other words, there is no ecdysis and such a stase exists only as a pharate stage *sensu* Hinton (1973). Several examples are provided in the second part. This clearly demonstrates that stases—as instars—must be defined with reference to apolysis and not to ecdysis (Coineau 1974:101). However, in some spiders and insects, calyptostases are able to emerge and thus undergo a proper ecdysis.[8] These special cases, unknown to Grandjean, necessitate the distinction of forms which undergo a proper ecdysis—to be called ectostases—from those which do not, endostases. Table 6.1 gives some examples of endo- and ectostases.

Table 6.1. Examples of States Observed in Stases

	Normal	*Elattostasis*	*Calyptostasis*
Ectostasis	Most cases	• Phoretic hypopus (deuteronymph) • Prelarva of Adamystidae • Larva of Labidostommidae • Decticous pupa in Megaloptera	• Prelarvae in *Diplura*, some Orthoptera and some Odonotaptera • Protonymph of *Lebertia* (*Hydrachnella*) • Numerous pupae in insects (adecticous pupae)
Endostasis	Unknown	• Adult of *Prosimulium ursinum* (in part)	• Most prelarvae in mites • Pupa in the Diptera-Cyclorrhapha

It must be emphasized that the state concept is essentially different from the stase and does not correspond to any level of ontogeny. The state only describes the appearance under which a stase can exist and there is no one-to-one correspondence between the two systems. For instance, a protonymph in mites may be calyptostatic, elattostatic or normal (i.e., walking and feeding); conversely, the calyptostatic inhibition may occur at any level of the mite ontogeny, from the prelarva to the tritonymph (see, fig. 6.2 below). Therefore, the two terminologies should not be confused as is often the case. This point is discussed to some extent by André and Jocqué (1986).

Homeomorphism and homostases

During the course of its ontogeny, or at least during a part of it, an arthropod may keep the same general appearance. This is generally the case of immatures in Collembola, nymphs in oribatid mites or caterpillars in Lepidoptera. These successive forms are real stases as demonstrated by a comparative study of chaetotaxy. Nevertheless, they are similar enough to be identified to species level whatever the stase observed. Such stases are said to be homeomorphic and are called homostases (Grandjean 1954); the sequence of successive homostases is sometimes called a phase (Vachon 1953; Grandjean 1954; Selander and Mathieu 1964; Hammen 1975).

In contrast, one stase of a species may differ markedly from the others.

This is the case of adult oribatid mites, hypopods of Acaridida, butterflies, and triungulin larvae, etc. Heteromorphism of oribatid adults is a major problem for soil zoologists who have to learn to recognize two quite distinct forms (immatures and the adult) for each oribatid species. Similarly, textbooks on acarology (e.g., Krantz 1978; Evans et al. 1985) provide two keys to identify Acaridida, one for the hypopods and another for the other stases. Such stases are said to be heteromorphic and are called heterostases.

There are no general and objective rules for determining whether two stases are homeo- or heteromorphic, because there is no fixed threshold level of similarity below which two stases could be considered heteromorphic. It is nevertheless indicated through the necessity of using different identification keys for different stases of a species. However, the criteria can vary to a large extent depending on the groups. For instance, in Collembola the first stase is usually considered to be heteromorphic (André 1987) which could surprise entomologists used to working on other insects. Actually, if stases are a matter of fact, phases are one of convention.

AGE-DEPENDENT EVOLUTION

The theory of age-dependent evolution is derived from the stase concept which was elaborated from observations of ontogeny in mites. This is why some illustrative examples of the diversity of mite development will be presented before I give an account of the theory itself. A few examples selected from other arthropods will support an extension of the theory to all arthropods, as suggested earlier. After this, I will illustrate how the theory alters our whole outlook on the relations between ontogeny and phylogeny, and I will also consider the biogenetic law. A simple but original example of formalization, the ontogenetic trajectory, will be made in the light of this theory. Such formalizations are a prerequisite for the incorporation of the theory into systematics.

Ontogenies in mite

Using the three described states (calyptostasis, elattostasis, and "normal" state), the ontogenies of different mite groups can be reexamined. Three examples were propounded by Grandjean (1957) and Hammen (1964)

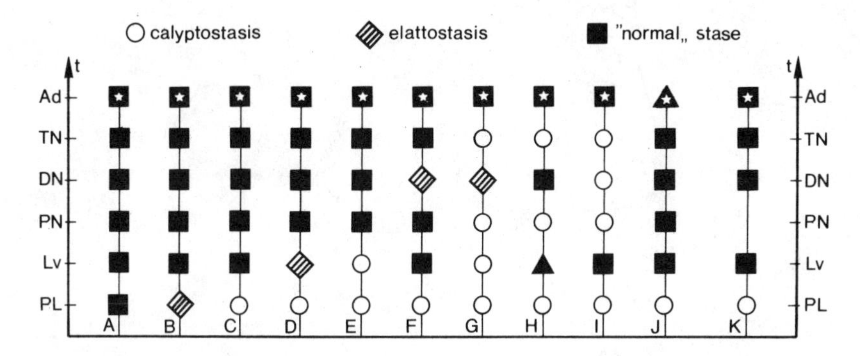

Figure 6.2. Ontogeny in mites. Different symbols, such as triangles designate heteromorphic stases (heterostases). Sexual maturity is indicated by white stars. A: archetype; B: Saxidromidae; C: Caeculidae; D: Labidostommidae; E: Myobiidae; F: Acaridae; G: Hypoderidae (*Hypodectes propus*); H: Erythraeidae; I: Ereynetidae Speleognathinae (*Boydaia nigra*); J: Oribatida; K: Tetranychidae. (Ad: adult; TN: tritonymph; DN: deuteronymph; PN: protonymph; Lv: larva; PL: prelarva; t: ontogenetic time).

to support their ideas. The first one refers to caeculid mites. The mobile forms observed in Caeculidae comprise one six-legged larva, three successive nymphs (proto-, deutero- and tritonymph) and the adult. These are stases, the chaetotaxy of which varies according to particular rules described in detail by Coineau (1974). Nevertheless, these stases are homeomorphic. In figure 6.2C, the five homostases *sensu* Grandjean (1954) are represented by the same black square symbol. The five stases are preceded by a calyptostatic prelarva which does not emerge from the egg which is represented by an open circle in figure 6.2. The ontogeny of *Tyroglyphus*, an acaridid mite, commences with a calyptostatic prelarva (fig. 6.2F). However, between the proto- and the tritonymph, a heterostase occurs called the hypopus. This dispersal form, frequent in Acaridida, is quite different from the other mobile stases. As the hypopods have no mouthparts and a reduced digestive tract (Boczek et al. 1969), they are unable to feed and are considered to be elattostases (a hatched diamond in fig. 6.2F). The third example selected by Grandjean (1957) and Hammen (1964) concerns the erythraeid mite *Balaustium*. Its ontogeny comprises the same number of stases as in the previous two species, but three of them—the prelarva, protonymph, and tritonymph—are calyptostases. The other three stases are mobile (figure 6.3). The deuter-

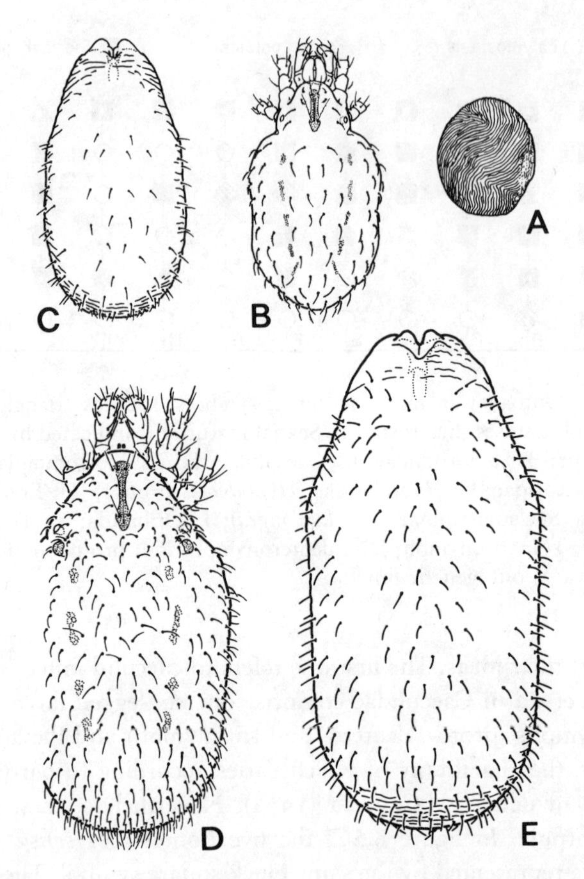

Figure 6.3. Development of *Balaustium florale*. A: calyptostatic prelarva; B: larva; C: calyptostatic protonymph; D: deuteronymph; E: calyptostatic tritonymph (from Hammen 1978; redrawn from Grandjean, 1957, 1959).

onymph and adult are homeomorphic. However, the larva is a heteromorphic parasitic form, with mouthparts quite different from those of other free-living forms (Witte 1978) (fig. 6.2H). A similar ontogeny had been described in a trombidiid mite by Henking (1882) and was also observed in water mites (Walter 1920; Cassagne-Méjean 1967).

Some other ontogenies are illustrated in Fig. 6.2. For example, Labidostommidae (Grandjean 1942) and Rhagidiidae (Grandjean 1945), have an elattostasis succeeding the calyptostatic prelarva (fig. 6.2D). However, the most unusual ontogenies are encountered in parasitic species. In some

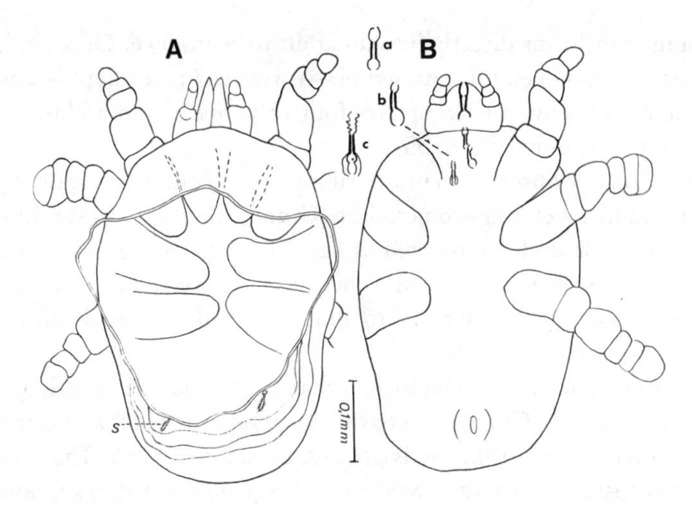

Figure 6.4. Larval exuvia of *Boydaia nigra* in dorsal (A) and ventral (B) views showing the pharynxes of the calyptostatic protonymph (a), deuteronymph (b), and tritonymph (c) (from Fain, 1972).

ectoparasitic Myobiidae, there are two successive calyptostases at the beginning of the ontogeny, and the first active stase, which is six-legged, might be a protonymph and not a larva (Grandjean 1938a) (fig. 6.2E). *Boydaia nigra,* an ereynetid mite living in the nostrils of birds, presents only two motile stases, the larva and the adult. However, a careful examination of the larva in pupation and of its exuvia reveals that the developing adult is not the only stase present within the larval skin. Indeed, it is possible to distinguish in the forepart of the pupa and exuvia some small "tubes" which represent the cuticular lining of the stomodeum and the pharynx of the "missing" stases (Fain 1972) (fig. 6.4). Actually, all the nymphal stases are reduced to calyptostases.

Hypodectes propus, an acaridid mite parasite of pigeons provides another example of extravagant ontogeny (Fain and Bafort 1967; Fain 1967). The hypopus emerges directly from the egg—i.e., it is an elattostatic deuteronymph. However, a careful examination of the egg containing the developing hypopus reveals the presence of the prelarva, larva, and protonymph, which are all reduced to an apoderma. As soon as it leaves the egg, the hypopus enters a young pigeon and stays in the subcutaneous fatty tissues until it leaves the host. Back in the nest, the hypopus transforms directly into an adult to reproduce. Once again, the "missing" stase—i.e., the tritonymph—is reduced to a calyptostasis. The complete ontogeny thus comprises four calyptostases in addition to the elattostatic hypopus (fig. 6.2G).

The ontogeny observed in most oribatid mites comprises a calyptostatic prelarva and a heteromorphic adult (Grandjean 1940, 1962; Sitnikova 1960). This life cycle, illustrated in figure 6.2J, is important for it shows that no intervening resting form is necessary between immatures and the heteromorphic adult, contrary to that observed in the so-called holometabolous insects.

It turns out, as shown in figure 6.2, that most prelarvae are calyptostatic in mites. However, Grandjean (1954a, 1962) speculated that the primitive prelarva had to be normal active stases, as in figure 6.2A. The discovery of an elattostatic prelarva in Saxidromidae (Coineau 1979) supports this theory (fig. 6.2B).

Another conclusion is that the number of stases is remarkably constant throughout the Actinotrichida, whether they are free-living or parasitic. The only case where a stase seems to have disappeared is provided by spider mites (Tetranychidae). Indeed, no apodermal skin has been found in either pupa intervening between the mobile stages (Van Impe 1985). Comparative chaetotaxy suggests that the missing stase could be the protonymph (fig. 6.2K).

The theory of age-dependent evolution

The major conclusion to be drawn from comparing the different life cycles in figure 6.2 concerns the occurrence of calyptostases. Calyptostatic inhibition occurs once or several times throughout the ontogeny, seemingly at random, between homeo- and heteromorphic stases. The only consistent explanation for calyptostasis is that this very regressive state manifests itself in a stase which followed its own evolution, irrespective of the stase which precedes or follows it in the course of development. As a calyptostasis is likely to occur at any level of the ontogeny, the previous statement implies that each stase—i.e., each level of ontogeny—follows its own evolution. In other words, *each stase has its own phylogeny*. This principle has been propounded by Grandjean (1947a) and is known as the "évolution selon l'âge" theory (1975b) translated as the theory of *age-dependent evolution*.

The principle defended by Grandjean does not mean that all the stases succeeding one another throughout ontogeny necessarily diverge from one another in their evolution. In numerous cases, the phylogenies, followed by the stases, parallel one another as if there were a coordinating system allowing the arthropods to evolve gradually and harmoniously from the first to the last stase. This evolution—characterized by parallelism of the level phylogenies, and ontogenetic concordance in the evolution of stases—is said to be harmonic (Grandjean 1947b; Hammen 1980). However, in other cases, stases evolve more or less independently from one another and give rise to several phases of development which differ morphologically and ecologically. Extreme evolutions, characterized by the divergence of level phylogenies, are said to be disharmonic (see below).

This theory, advanced by Grandjean, conforms well with the one elaborated previously by Garstang (1922). The level of ontogeny defined by Grandjean corresponds to Garstang's grade of differentiation, and his theory is well summarized by statements VI and VII, propounded in Garstang (1922):

> VI. As the individual, through all the form-changes of his life-cycle, is an evolutional and functional unity, modifications manifested in his larval or adult phases involve co-ordinating changes in the more passive and formative phases (embryonic, post-larval, pupal stages).
> VII. Thus, while a given ontogeny, under normal conditions, tends to repeat the form-sequences of its predecessors, it is liable to changes in every part of the life-cycle—positively, by equipping the larval and adult stages for the changing conditions of their various careers, or with greater efficiency for the same conditions, and negatively, by abbreviating the formative processes to the uttermost.

Garstang's statement VI corresponds to harmonic evolution of Grandjean while statement VII describes disharmonic evolution. However, these principles are still misunderstood. Cohen and Massey's (1983) definition of Garstang's system as a linear developmental model is a good example of misinterpretation. The two theories have altered our whole outlook on evolutionary biology. They offer a nice example of convergent theories since one is based mainly on the observations of marine organisms such as the Tunicata, and the other is based on a study of mites. Nevertheless, the latter is original in that it is based on animals showing a discontinuous ontogeny (the stases) and is general in that it can be applied to all arthropods, especially to insects.[9]

Ontogenies in other arthropods

As noted by Hennig (1969, 1981), very little is known about the stages in insects, the information available from the literature being extremely limited. Indeed, most ontogenies in the literature consist of a brief description of the different instars. In numerous cases, the description is even restricted to the last larval instar. The information concerning the number of instars is often vague and that related to the character transformation through ontogeny is seldom published. In addition, endostases are often not recorded since, by definition, they are not immediately apparent or give rise to contests as for instance in the so-called fourth-stage larva of tsetse flies which precedes the true pupa and remains within the puparium (Fraenkel and Blashkaran 1973). Calyptostases which do not leave the egg are also difficult to record; for instance, the calyptostatic prelarva of Lepidoptera was observed relatively recently by Okada (1958). The biological significance of the later calyptostases is often overlooked and, even when known for a long time, are not mentioned in some textbooks. In part, this explains why so few examples are given here. Furthermore, the accuracy of the chosen examples cannot be guaranteed, because of the present state of our knowledge. Neither can it be assumed that the examples are the only ones, or even the best ones to illustrate the point being made.

Different examples relating to spiders and other nonacarine arachnids are given in André and Jocqué (1986). Briefly, the number of stases is well-defined in such groups as false scorpions, which exhibit six stases, but seems to be undefined, at least at first sight, in spiders. The number of calyptostases found at the beginning of ontogeny in spiders varies from one to three depending on the family and the calyptostases are usually followed by an elattostasis.

In insects, Collembola offer an illustrative example of the difference between the number of stases and the number of instars (André 1987). The number of instars is very often undefined and very high (up to 50 molts); it is said to be n or $(6 + n)$ if there are six immature instars. However, the number of stases is well-defined. In *Folsomia quadrioculata*, for instance, the number of stases defined from the genital chaetotaxy should be eight, although sexual maturity is reached after five ecdyses (Grégoire-Wibo 1974; fig. 6.5A) and the total number of instars is much

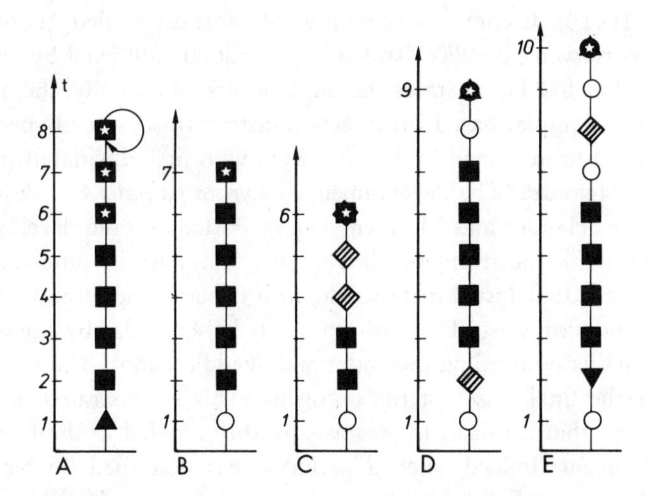

Figure 6.5. Ontogeny in insects. Same conventions and symbols as in fig. 6.2. A: Collembola (*Folsomia quadrioculata*); B: Orthoptera; C: Thysanoptera Tubulifera; D: Cantharidae; E: Meloidae (*Pyrota palpalis*).

greater than eight. After the eighth stase, all further instars are merely self-replicates. Other examples of ontogenies in Collembola are given by André (1987).

Ontogeny in Orthoptera begins with a calyptostasis described in the last century (Kunckel d'Herculais 1890). As early as 1918, its existence was well interpreted by La Baume and then by Uvarov (1928), who called it a *vermiform larva*. The following stases, called hoppers, are active feeding stases. They differ from one another by various characters, such as the number of antennal segments (Roonwall 1952; Uvarov 1966). The number of hoppers is usually five, but varies depending on the species (fig. 6.5B).

Figure 6.5C refers to Thysanoptera. Once again, the ontogeny starts with a calyptostatic prelarva (Lewis 1973). This prelarva is followed by two active stases, usually called larvae, which ingest all the food necessary for development to the adult. In Tubulifera, two resting stases follow the larvae; they are elattostases as they do not feed but can walk slowly if disturbed. At the final molt, the adult emerges from the pupa.

Figure 6.5D refers to the development of Cantharidae as described by

Verhoeff (1919). It commences with a calyptostasis (called "Fotus-studien," "Vorlarven" or "Fötalstufen" by Verhoeff) followed by an elattostasis. After five larval stases the pupa occurs and, finally, the adult.

The last example, based on insects, illustrates a case of hypermetamorphosis, a term coined by Fabre (1857) who first elucidated the ontogeny of Meloidea. The development of *Pyrota palpalis* was described in detail by Selander and Mathieu (1964). Postembryonic development is initiated with the triungulin larva. This first stase is quite different from the next four larval instars. The fifth instar is then followed by a calyptostatic sixth stase. The sixth stase may be succeeded by a new grub larva, which is nonfeeding (Selander and Weddle 1969). The pupa and adult are the final stases of the ontogeny. This is illustrated in figure 6.5E, except that a third calyptostasis has been added at the first level of development. Indeed, such a prelarva was described by Newport (1841), but it seems that his observation has been completely forgotten by modern authors. Numerous variants of development are known in Meloidea depending on the species (Selander and Mathieu 1964) or temperature (Selander and Weddle 1969).

These few examples clearly illustrate that the stase concept and the derived theory of age-dependent evolution can be extended to insects. Extension to myriapods should not pose any problem although the terminology of development is somewhat confusing, and the boundary between juvenile and adult stases seems to be difficult to define correctly (Anderson 1976). Calyptostases are known to occur at the beginning of the ontogeny (Verhoeff 1905). The development of myriapods is characterized by anamorphosis, where segments and segmental appendages are added during postembryonic development (Verhoeff 1905).

As in myriapods, anamorphosis is known to occur in Crustacea. They provide a wide range of ontogenies extensively discussed in Jägersten (1972). A typical ontogeny is found in cyclopoid copepods; this entails four phases involving two naupliar, four metanaupliar, six copepodid and one adult stases. As in insects and mites, parasitism induces peculiar ontogenies. In parasitic copepods, the naupliar and metanaupliar stases may be passed within the egg (Matsuda 1979) and are then considered as calyptostatic endostases. This clearly confirms that the evolutionary mechanisms involved in crustaceans are similar to those found in insects and mites, and that the stases concept and the related principles may be extended to arthropods as a whole.

Relations between ontogeny and phylogeny

The major strength of Grandjean's work lies in his effort to formalize his theory from the comparative study of ontogenies. An example of such a formalization is provided by what he called *diagrammes chronologiques* (Grandjean 1947b; Hammen 1964). Although Grandjean's diagrams are similar to those elaborated by de Beer (1930), they differ from them in two points. First, they refer to stases, i.e., they deal with discontinuous ontogenies; second, they refer to "all or none" characters. André (1979) slightly modified Grandjean's diagrams and renamed them ontophylogenetic diagrams.

To describe the ontophylogenetic diagrams let *P* be the primitive (plesiomorphic) state of a character—i.e., a state which through phylogenetic time, *T*, precedes a derived or secondary (apomorphic) state *S*. States *P* and *S* are exclusive and correspond to discrete data. In a 2–state character, the system is thus represented by binary data, e.g. presence and absence data. What happens through ontogenetic time, *t*? A priori, there are three possibilities: *P* precedes *S*, or *P* comes after *S*, or thirdly either *P* or *S* may be present throughout the entire ontogeny. The same three situations are conceivable with 3–state characters, provided that the states are exclusive, correspond to discrete data, and form a linear phylogenetic sequence whose polarity is determined and fixed. Actually, this may be generalized to any *n*-state character meeting the three conditions cited.

The three situations are found among tydeid mites (André 1979) and are illustrated by ontophylogenetic figures (fig. 6.6, A-C). Figure 6.6A illustrates a 3–state character and refers to the tarsal setae *(it)* of leg *I* which are sometimes eupathidial. The eupathidial character is considered to be the most primitive state, while the normal seta is the derived state in the family; the absence of setae is a character which is still more advanced. These three states are plotted against ontogenetic time *t*, and phylogenetic time, *T*. Lines may be drawn to separate the three states. Plotting this data results in what Grandjean (1947b, 1951b) called an ascendent harmony or, since it is a regression along time *T*, a proregression. This situation corresponds to neoteny (*sensu* Kollmann 1885) or to the retardation process of de Beer (1940). Through ontogeny, the character *S* precedes character *P* or, briefly, the larva is "more advanced" than the adult in this respect. The derived character extends into adjoining

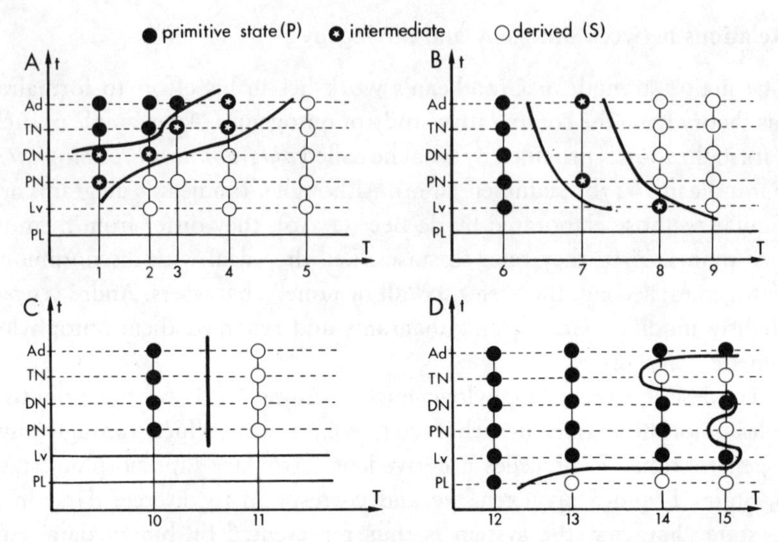

Figure 6.6. Ontophylogenetic diagrams showing ascendant harmony (A), descendent (B) and vertical harmony (C) and disharmony (D). A, B, and C refer to examples observed in Tydeidae (after André, 1979). A: iteral setae are eupathidial (primitive state), normal (intermediate) or absent (derived). B: apotele I normal (primitive), reduced with only the empodium (intermediate) or absent (derived). C: protonymphal genital acetabula present (primitive) or absent (derived). D refers to the calyptostatic inhibition (derived character) occurring at different levels in different families. (t: ontogenetic time; T: phylogenetic time; 1–11: tydeid species (listed in André 1979); 12: archetype; 13–15: different families of mites.) Stases that are unknown in some species are not indicated in diagrams.

stases from the larva to the adult. Note that in diagram 6.6A, the prelarval level is not considered since it is calyptostatic in all Tydeidae. "Negative" neoteny—i.e., the absence in the adult of a character which should have appeared during the ontogeny,—is very common in Acari (Grandjean 1938b) and is also described in Opiliones (Rambla 1980).

Figure 6.6B refers to another 3–state character. *P* is represented by the presence of apotele I, the derived state is the presence of vestigial claws[10] and the very advanced state is the complete disappearance of the apotele. The resulting figure B is the reverse of figure A, in that the separating lines are descendent. This is a descendent harmony or, since it is a regression along time *T*, a retroregression. This corresponds to tachygenesis *sensu* Garstang (1928) or acceleration *sensu* de Beer (1958). The derived character extends into adjoining stases backward from the adult toward the larva, which is "less advanced" than the adult in this respect. The latter example produces a parallelism (recapitulation) between ontogeny and phylogeny, unlike the former example.

Figure 6.6C refers to the third possibility and deals with the protonymphal genital acetabula. These genital acetabula are either formed in the protonymph (the larva is at a deficiency level in this respect, as is the prelarva) and are present through the entire ontogeny, or they do not appear in the protonymph or in any of the later stases. This is called a *vertical harmony* or, since it is a regression, a vertical regression.

Grandjean was impressed by the fact that for a given character, the ontogeny of a species, or the ontogenies of a group, are cut only once by the line *PS* (i.e., the line separating character *P* from *S* in figure 6.6A-C). An exception occurs however when there is a disharmony resulting from the presence of calyptostases. Such disharmony is illustrated in figure 6.6D, which refers to the presence of apotele *I* in different mite families; in such a case, the line *PS* is broken or curved.

If several ontogenies are studied along time *T* and if there is no disharmony, then the harmony remains of the same type. Hence, the principle of concordance in harmonic evolution which states that, if a change occurs once through any ontogeny, then it occurs only in one way in the considered group.

Another conclusion to draw from a study of figures 6.6A-C concerns the distinction between amphi- and eustasy. A character is said to be eustasic when it always appears at the same level in the course of ontogeny; when it does not appear at that level, it is absent at all levels (Grandjean 1958; Hammen 1980). Genital acetabula are eustasic in actinotrichid mites as are the trichobothria on the chelae of false scorpions (Vachon 1934). In contrast, some characters are amphistasic and, depending on the species, can appear at any level of the ontogeny. For instance, the base level for the appearance of the iteral setae on tarsus *I* of Tydeidae is the protonymph, but in some species their appearance is delayed until the deutero- or tritonymphal level (André 1981b). In the Phytoseiidae and Ameroseiidae (Mesostigmata), sternal setae (genital setae) appear in the deuteronymph instead of in the protonymph as in other Gamasina (Evans and Till 1979). Numerous other examples of amphystasy are known in mites.

The biogenetic law and related principles

Figure 6.6 clearly illustrates that the biogenetic law does not apply to postembryonic development of arthropods. Indeed, if recapitulation is supported in the case of descendent harmony, it is refuted by ascendent harmony. The latter case is related to neoteny, defined as the retention of young features or the retardation of somatic development (Gould 1977:228). Neoteny is one of the two falsifiers of the biogenetic law as discussed by Nelson (1978) and Nelson and Platnick (1981). However, both authors conclude that neoteny could be an apparent falsifier and further suggest that it would be a reflection of a lack of information rather than a real falsifier.

The other falsifier of the biogenetic law as discussed by Nelson (1978) and Nelson and Platnick (1981) is the situation of contradictory ontogenetic character transformations. Unfortunately, both authors admit to being unaware of any examples of it. Yet, such an example does exist and is provided by Oribatida. Most oribatid mites have a pair of trichobothria located on the prodorsum which are composed of a sensillus, often clublike, arising from a large bothridium. This character, found also in other actinotrichid mites, may be reasonably considered as the ancestral or primitive state in Oribatida. In different families of Oribatida, particularly in the genus *Camisia,* the sensillus is reduced to a minute seta totally without or with a small bothridium in the immature form. This character is thus neanic *sensu* de Beer (1958). This "advanced" state is found in most, if not all, *Camisia* larvae and depending on the species, extends into adjoining stases from the larva to the adult. Our knowledge of the genus is good enough to draw an ontophylogenetic diagram covering the near complete range of developments (fig. 6.7A). This *Camisia*-type trichobothrial regression, first described by Grandjean (1939), is another example of ascendent harmony or neoteny which also contradicts the biogenetic law. The most interesting point is that the contradictory trichobothridial transformation has been discovered and named *Hydrozetes*-type trichobothridial regression by Grandjean (1951a). In the oribatid genera *Hydrozetes* and *Limnozetes,* the sensillus in immatures is setiform and arises from a large bothridium while in the adult of some species the bothridium and sensillus are both reduced (fig. 6.7B). Thus the reduction is ephebic *sensu* de Beer (1958). This regression is an example of descendent harmony and is much rarer than the *Camisia*-

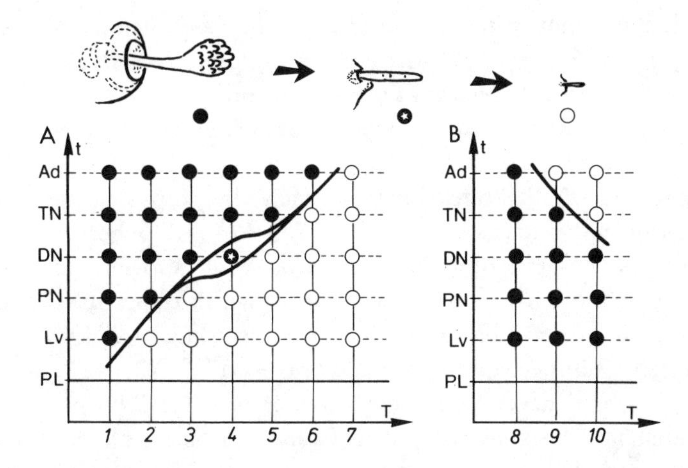

Figure 6.7. Ontophylogenetic diagrams showing the *Camisia*-type (A) vs. the *Hydrozetes*-type trichobothridial regression (B). Symbols and conventions as in fig. 6.6. The upper part of the figure illustrates the *Camisia*-type regression from the clublike sensillus arising from a large bothridia (primitive state) to the minute sensillus (derived state) in *Camisia carrolli* (figures 1, 7 and 10 refer to theoretical cases, i.e. not yet observed; 2–6: different *Camisia* species; 8–9: *Hydrozetes* species) (from André et al. 1984).

type trichobothrial regression. Both types of regression together provide a good example of contradictory ontogenetic character transformations and constitute a real falsifier of the biogenetic law as stated by Nelson (1978) and Nelson and Platnick (1981). Indeed, whichever character state is selected as primitive (i.e., whichever polarity), the same conclusion must be drawn.

The law does not hold up, irrespective of its formulation. Nelson (1978) restated the biogenetic law as follows: *given an ontogenetic character transformation, from a character observed to be more general to a character to be less general, the more general character is primitive and the less general advanced.* Because it doesn't apply to some groups, as I have demonstrated in the previous examples, Nelson's restatement has also been criticized by Voorzanger and Steen (1982), Alberch (1985), Brooks and Wiley (1985), de Queiroz (1985) and Kluge (1985). See Nelson (1985) for his response.[11]

Another development is the theorem proposed by Løvtrup (1974) which states that *in the course of their ontogeny the members of a set of twin taxa follow the same course of recapitulation up to the stage of their divergence into separate taxa.* This theorem was restated in a slightly modified form by Wiley (1981:156) who related it to *Baerian recapitulation.*

Wiley (1981) illustrated Løvtrup's views by proposing linear ontogenetic sequences such as clade $A{:}Y^1 \rightarrow Y^2 \rightarrow Y^3 \rightarrow Y^4$ where *Y*'s represent the successive states of a character throughout ontogeny. He then supposes that the gene, or the gene complex which transcribed to give Y^3, mutates and transcribes to produce Y^7. According to Wiley (1981), the expected resulting sequence is clade $B{:}Y^1 \rightarrow Y^2 \rightarrow Y^7 \rightarrow Y^8$ and he adds that he would never expect the ancestor of clade *B* to have an adult terminal Y^4. This view contradicts Grandjean's principle that each stase has its own phylogeny, irrespective of that of others. It is refuted by the occurrence of calyptostases at any level of ontogeny.

Another example using Wiley's developmental sequences could be two clades, C and *H,* which show the primitive (plesiomorphic) state of a character at each level of their ontogeny—for instance, a normal trichobothria. If the primitive state is coded *Y,* we have clade $H{:}Y^1 \rightarrow Y^1 \rightarrow Y^1 \rightarrow Y^1$ and clade $C{:}Y^1 \rightarrow Y^1 \rightarrow Y^1 \rightarrow Y^1$.

From the common ancestral sequence, two different sequences may be derived: clade $H'{:} \rightarrow Y^1 \rightarrow Y^1 \rightarrow Y^2 \rightarrow Y^3$ clade $C'{:}Y^3 \rightarrow Y^2 \rightarrow Y^1 \rightarrow Y^1$ where Y^3 represents the most advanced state (for instance, a reduced trichobothria and Y^2 an intermediate state (as in fig. 6.7). If the two ancestral species C and *H* survive, and the ancestral sequence is kept in those species, the outcome is two groups of sister taxa, *H* and *H' vs.* C and *C'*. In the former case, the two species follow the same course of recapitulation; in the latter, they do not. Similarly, there is a gradual change from the general to the special in the first case, but the converse in the second. The first case is illustrated by the *Hydrozetes*-type regression and the second by the *Camisia*-type regression already mentioned above. This clearly refutes the theorem advanced by Løvtrup (1974).

Løvtrup's theorem is denied also by the practical experience of systematists and ecologists who work with both immature and adult forms since, it happens that immature forms of two species are much easier to distinguish from one another than the adults which look like sibling species. This observation corresponds to Giard's (1905) concept of poecilogony. An illustrative example is offered by the oribatid genus *Ori-*

batella, one species of which lives in the soil and another on bark. The larvae are more easily distinguished than the adults. Other examples in Oribatida are commented upon by Travé (1964). Another example is offered by the Simuliidae, particularly by the *Simulium damnosum* complex, where the adult stase is the most difficult to identify in comparison to larvae (Meredith 1984). As a result, descriptions of new species are based on larvae (e.g., Elsen et al. 1983). All these examples, and many others, show that divergence in characters appears at all stases in ontogeny. They deny that, in the course of ontogeny, there is necessarily a gradual change from the general to the special.

Canons and ontogenetic trajectories

Evolution cannot be considered as a random phenomenon; on the contrary, it may be supposed to be governed by laws and to follow a hierarchical type of strategy (Koestler 1972).

Koestler (1967) developed the general properties of an element of such a hierarchy—called holon by Koestler or integron by Jacob (1970). The first two are:

1. Functional holons are governed by fixed sets of rules and display more or less flexible strategies.

2. The rules—referred to as the system's *canon*—determine its invariant properties, its structure, configuration, or pattern.

In this context, stases are holons and ontogeny in mites and other arthopods obey *canons,* laws which guide their evolution. In logic, such canons are expressed in terms of relations. A relation between sets A_1, $A_2, \ldots, A_n$ is a subset of the cartesian product $A_1 \times A_2 \times \ldots \times A_n$: $R_{1 \leq i \leq n\ (At)} \subseteq X_{1 \leq i \leq n} A_i$ A good example of such a relation is offered by genital and aggenital setae in Tydeidae. Set A_1 comprises six aggenital elements as it includes five aggenital setae plus their possible common absence. Set A_2 comprises six genital setae, which make seven elements if their absence is considered. If ontogeny has to be considered, a third set, A_3, comprising the five active stases, must be taken into account. Thus, the cartesian product involves 210 possible combinations, among which only 24 have been recorded. These 24 combinations represent operational semaphorontic units (OSUs) (i.e., they correspond to semaphoronts such as the male adult, the female tritonymph) and are plotted in a 3–dimensional diagram in figure 6.8A. The successive combinations

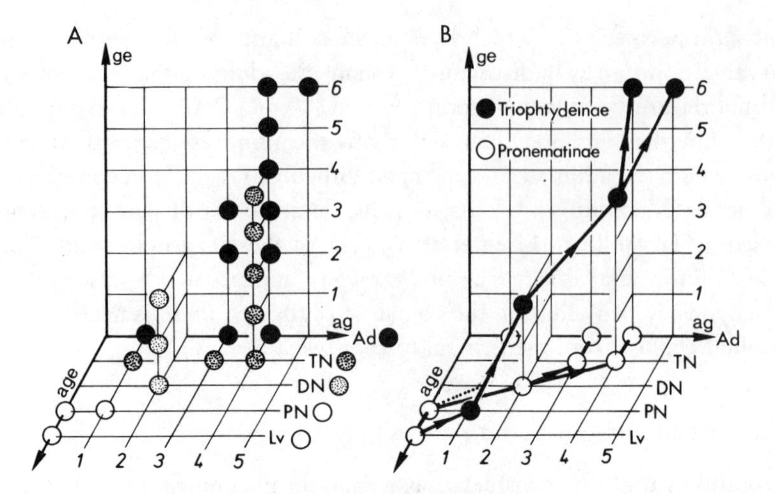

Figure 6.8. A Diagram illustrating the relationship between the number of genital (age) and aggenital (ag) setae in the active stases of Tydeidae. Ontogenetic trajectories drawn from data of diagram A for the subfamilies Triophtydeinae and Pronematinae. Stase identification as in fig. 6.2.

encountered through a given ontogeny form an *ontogenetic trajectory* in figure 6.8B. Such trajectories permit the discrimination of different subfamilies and clearly demonstrate that ontogeny in mites is not a random phenomenon, but obeys *canons,* which vary depending on the taxa and the taxonomic rank under consideration.

This simple example also shows that it is possible to go beyond the verbose description usually found in systematics. In the same way that growth may be formalized in mathematical terms (see, e.g., Thom 1977), development could be translated into a set of relations. Although a thorough study of such formalization is beyond the scope of this article, it must be admitted that they constitute general requirements for modeling developmental processes and incorporating them into systematics.

SYSTEMATIC IMPLICATIONS

In the previous two parts, the stase concept and the theory of age-dependent evolution have been presented. A few attempts in the past have been made to incorporate these aspects of development into systematics but these have met with resistance. For instance, Snodgrass (1954) wrote:

"Insects cannot be classified according to the type of metamorphosis they undergo." As early as 1873, the same statement was made by Girard: "Les métamorphoses ne sont pas l'indice d'une distinction fondamentale, puisque toutes les différences s'expliquent par des arrêts de développement; elles ne donnent donc que des médiocres charactères pour une classification philosophique." The idea that metamorphosis in insects is essentially a prolongation of the embryonic stage is an old one (Lacordaire 1834; Girard 1873), but is still defended by some biologists (Brien 1974; Matsuda 1976). This idea is in opposition with the stase concept advanced by Grandjean, who made a clear distinction between the embryo and the subsequent postembryonic stages (1940). The subsequent assertion that insects or other arthropods cannot be classified according to their development clearly disagrees with Garstang's principle that ontogeny creates phylogeny.

If ontogeny creates phylogeny, as claimed by Garstang, while the classification of organisms has to reflect the genealogical relationships among them, it would be inappropriate to omit ontogenies from systematics. This view was articulated by Danser (1950) who claimed that systematists never classify objects, but life cycles; and he concluded that the life cycle, with its multiformity, is the smallest unit of classification. This view also underlies Hennig's (1950) work, which refers to the hologeny[12] concept proposed by Zimmerman (1948) and which is still defended by Leppik (1970). However, a major problem arises because there is no real theory to explain the evolution of ontogenies, except a few systems such as those of Hammen (1974, 1978, 1979) and Jägersten (1972). Hammen's system will be considered in greater detail below. Such an approach has had some applications in major classifications (e.g., at ordinal or familial level) and is based mainly on the phenomena of heterostasy and elassostasy, as will also be explained below.

At a lower taxonomic level, most classifications of arthropods are based entirely on adults. The persisting prejudice that only mature individuals are important for classification is derived from the Aristotelian notion that individuals do not fully manifest the essence (= form, purpose) of the species until they reach maturity (Griffiths 1974). However, there are exceptions, such as chigger mites where the classification is based on larvae. In that group, the importance of larva has been stressed to such an extent that it is considered to be the "repository" of phylogeny and taxonomy by Vercammen-Grandjean (1969a), who later introduced the term nepophylogeny (1969b, 1973). In Acaridida (Astigmata), the classification of several families has relied to a great extent on the hypopods—i.e., the phoretic deuteronymphs.

In some groups, different classifications for the larvae and the adults exist. In such cases, the problem of incongruencies between larval and adult groupings arises. The problem of congruency in insects has been investigated early by Lenz (1926) and Emden (1927, 1957). According to Emden (1957), genuine incongruities between both classifications are rare and indicate that one of the systems concerned is unnatural—i.e., phylogenetically wrong. However, even classifications of this type, whether based on the larvae or adult, rely on an arbitrary selection of an ontogenetic level, and quite disregard the development itself—i.e., the dynamic process leading from the first to the last stase, from birth to reproduction. Yet, considering an animal throughout its ontogeny—from the first to the last stase—is the only way to analyze it as a functional and developing unit, and to identify the different characters succeeding one another. In this context, I disagree with de Queiroz's (1985) statements that characters do not transform in ontogeny and that ontogenetic transformations themselves are the characters. To incorporate ontogeny into systematics, two original approaches will be presented below: ontogenetic trajectories and ontogenetic dendrograms. Both techniques rely on an understanding of heterochrony.

Evolution of ontogenies

If ontogeny creates phylogeny and ontogeny has to be considered in systematic research, then the evolution of ontogenies has to be understood. Yet only typologies of development exist, as for instance that proposed by Weber (1933). Although widely used (e.g., Joly 1977), Weber's typology has undergone many modifications and adaptations, depending on the authors. In addition, Weber's system relies on the recognition of different phases (larva, pupa, etc.), which is inconsistent with the stase approach (André and Jocqué 1986). As a result, the terminologies referring to "special" ontogenies are somewhat puzzling. The term "hypermetamorphosis" was coined by Fabre in 1857, whereas Snodgrass (1954) and Chapman (1969) prefer the term "heteromorphosis." The expression "complex life cycle" is used by Istock (1967) and abbreviated to CLC by Wilbur (1980). Matsuda (1979) deals with "abnormal metamorphoses" and coins the term "halmatometamorphosis" for "excessive" metamorphoses occurring in parasitic arthropods. Even terms designating evolutionary processes are equivocal. The term "acceleration" as used by Matsuda (1979)[13] refers to the elimination or omission of large numbers of developmental stages, which disagree with the original meaning of de Beer (1958) related to heterochrony, or the meaning proposed later by Gould (1977).

Matsuda (1979), citing the example of ereynetid mites, states that acceleration occurs by omission of nymphal stages. This assertion is false since, as explained previously, all the stases exist in Ereynetidae but some of them are reduced to calyptostases. Fain (1969) uses the term "acceleration" for those ontogenies characterized by multiple calyptostases. Consequently, the concept of acceleration is ambiguous as it has been defined in two quite different ways (de Beer vs. Matsuda) and because it designates both the real omission of stases or their reduction to the calyptostatic state (Matsuda vs. Fain). Furthermore, it is misleading since, in numerous cases, it is used to designate the omission of stases, whereas the so-called "missing" stases exist as calyptostases.

The only consistent system based on the stase concept has been proposed by Hammen (1974, 1975, 1978, 1979). According to him, the evolution of the chelicerate life cycle started with stasoids and there was only one immature phase (stasoidy, fig. 6.9). It is supposed that stases originated from stasoids, which implies both the acquisition of the idionymic character and constancy in the number of molts (orthostasy) (central part, figure 6.9).

The primitive orthostasy comprises a given ancestral number of stases (e.g., five as in fig. 6.9) and is called protostasy by Hammen. In the course of evolution, the ancestral orthostasic life cycle has been subjected to different evolutionary phenomena, which are illustrated in figure 6.9. A first trend is heterochrony in the sense of de Beer (1930) (left-hand side, fig. 6.9). As analyzed by Gould (1977:228), heterochrony involves only two processes of acceleration (also called adultation by Jägersten 1972) and retardation. In the former case, a feature appears earlier in ontogeny of a descendant than it did in its ancestor. In retardation, the feature persists later in ontogeny than it did in its ancestor. However, two special cases must be stressed. The acceleration of sexual maturity in insects and other arthropods has been named prothetely (Sing-Pruthi 1924; Wigglesworth 1954; Matsuda 1979) or paedogenesis (Matsuda 1979), while

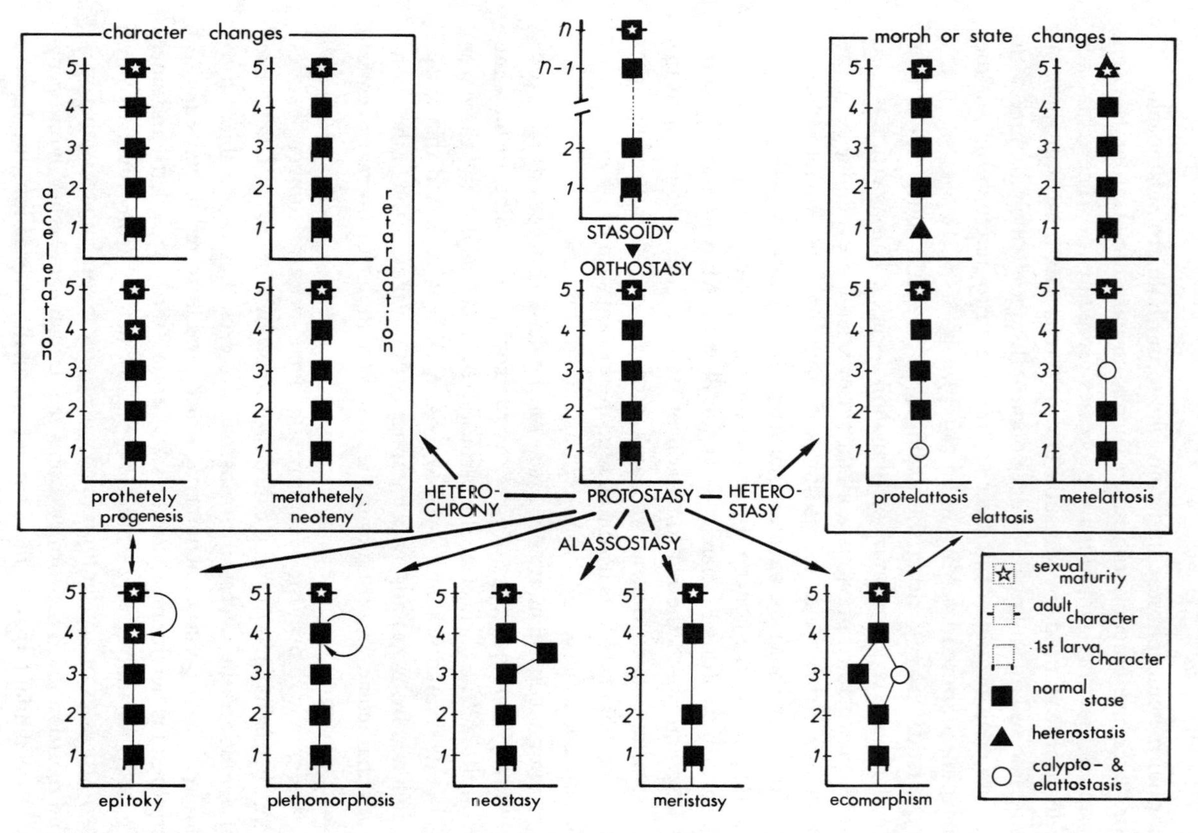

Figure 6.9. Evolution of ontogenies from stasoidy to orthostasy with subsequent evolutions derived from protostasy (see text).

the persistence of a larval character until the adult stase has been called metathely[14] (Wigglesworth 1954). In the latter, the arthropod becomes a neotenous adult after undergoing a normal number of molts; in the former, the arthropod becomes adult after undergoing less than the normal number of molts.

The right-hand side of figure 6.9 is concerned with heterostasy—i.e., the occurrence of heterostases in the course of development. Such heterostases may occur at any level of ontogeny, at the beginning (for instance, the triungulin larvae in Meloidae) or at the end (the adult stase in oribatid mites). As special cases of heterostasy, the occurrence of elatto- and calyptostases must be mentioned. The presence of calyptostases at the beginning of ontogeny seems to be a general phenomenon in arthropods and has been termed protelattosis by Hammen (1975, 1980). In contrast, the occurrence of intervening elatto- or calyptostases as distinct from those at the beginning has been called metalattosis. Good examples are offered by the elattostatic hypopus in acaridid mites, or by the calyptostatic pupa in holometabolous insects.

The last major trend (lower part, figure 6.9) is alassostasy. The term originally coined by Hammen (1975) is designated to any ontogeny characterized by secondary changes in the number of stases, neostasy, and meristasy. Meristasy—i.e., the disappearance of a stase—is by definition questionable, as the real absence of a stase is difficult to prove. In addition, without careful observation, meristasy can be easily confused with prothetely. The case of the missing nymph in Tetranychidae has already been mentioned. Neostasy—i.e., a secondary increase in the number of stases (Grandjean 1970)—is the only phenomenon not yet formally identified, as there are no criteria to distinguish neostases from stasoids (Hammen 1980). Neostasy should not be confused with plethomorphosis, which consists of a secondary formation of isophena at a certain level of postembryonic development (Hammen 1975, 1980). Isophena are known, for instance, in argasid ticks (Hammen 1980).

The final two cases illustrated in figure 6.9 have been recorded in Collembola. The first is epitoky (Bourgeois 1971; Bourgeois and Cassagnan 1973), which consists of the alternation of egg-producing and nonproductive female instars. Provided that the epitokous form differs from the other by "all or none" character(s), such as the presence/absence of the exsertile antennal sac (Bourgeois 1974), then epitoky implies the succession of two adult stases. The second, ecomorphism (Cassagnau

1956, 1972) relates to polymorphism rather than to polyeidism. It is, however, related to elattosis insofar as the ecomorphic state corresponds to an elatto- or calyptostasis. It differs from elattosis in that the elattostatic state is facultative and initiated by climatic conditions. Another example of ecomorphism is given by Meloidae in which temperature plays a critical role in the appearance of calyptostases (the so-called coarctate larva) instead of grubs (Selander and Weddle 1969).

The evolution of ontogenies are discernible in some cases. Evolution toward protelattosis is well known in spiders, where the number of calyptostases observed at the beginning of the life cycle varies from one, in primitive families, to three (André and Jocqué 1986). Passage from an ectostatic calyptostasis to an endostatic calyptostase is illustrated by *Hydrachnella,* where the larval exuvia may or may not be rejected, depending on the genus and species (Cassagne-Méjean 1963). A similar observation was made on the adult of *Prosimilium ursinum,* which is an elattostasis failing to emerge from the pupa in numerous, but not all, individuals (Carlsson 1962). The transformation from an elattostasis to a calyptostasis in mites may be reconstructed from the observations of Newell (1973) and Coineau (1974, 1979). A similar trend is observed in insects, where the pupa shows different degrees of inhibition from the *pupa dectica* to a mere apodermal skin (Hinton 1948; Hinton and Mackerras 1970).

However, convergence in ontogeny must be expected, like convergence in morphological characters. For instance, an ontogeny with three intervening calyptostases (as in fig. 6.2H) is known to occur in Erythraeidae and in Trombiculidae, two closely related families whose larvae are parasitic. The same ontogeny, however, is found in *Hydrachnella* (Cassagne-Méjean 1963) and even in Pterygosomatidae, which on the basis of morphological characters are usually classified as distinct from Erythraeoidea. Like morphological convergences, ontogenetic convergences make classification more difficult, but such difficulties should not prevent systematists from using ontogeny for systematic purposes, as was suggested by Snodgrass (1954).

Ontogeny and major classification

As noted by Snodgrass (1954), holometabolism is not limited to the group of orders formerly known as Holometabola, and even in orders that are typically ametabolous there may be juvenile changes during growth sufficient to warrant the term paurometabolism. Nevertheless, Apterygota and other primitive orders have simple ontogenies, possibly complicated by protelattosis (i.e., the occurrence of calyptostasis at the beginning of the life cycle). Metalattosis and 2– and 3–phase ontogenies are recorded only in more specialized orders of insects and mites. The fundamental number of stases in Acariformes (Actinotrichida) differs from that observed in Parasitiformes, and this difference is of major importance in characterizing the two major groups of mites (Lindquist 1984).

In 1946b, Hinton proposed a new subordinal classification of the Lepidoptera, and erected the suborder Dacnonypha. In addition, he transferred the Micropterygidae, a family usually placed in Lepidoptera, to a separate order, the Zeugloptera.[15] A key character used by Hinton was the presence of a decticous pupa in Zeugloptera, Trichoptera, and Dacnonypha compared with the presence of an adecticous pupa in other Lepidoptera. This character clearly refers to a degree of calyptostatic inhibition. Similarly, hypermetamorphic development is a key character used by Pierce (1964) for considering the Strepsiptera as a distinct order, while Kristensen (1981) concludes that since three calyptostatic stases precede the adult, even the assignment of the Strepsiptera to the Holometabola cannot be considered unequivocally.

Finally as illustrated in figure 6.8, different subfamilies of Tydeidae can be discriminated through genital chaetotaxy and related ontogenetic trajectories. In Ereynetidae, a mite family closely related to Tydeidae, Fain (1957) distinguished three subfamilies. As he later demonstrated (1963), each of them is characterized by a peculiar ontogeny: most Ereynetinae are free-living and have a simple calyptostatic stase, the prelarva. In Lawrencarinae, parasitic on batracians, the tritonymph is also a calyptostasis, whereas in Speleognathinae, the most involved subfamily parasitizing warm-blooded vertebrates, all three nymphs are reduced to calyptostases.

It is evident from the preceding account that the evolutionary processes observed in arthropod ontogeny as detailed above have to be considered in defining the major divisions in systematics.

Ontogenetic dendrograms and systematics

The incorporation of a dynamic process such as ontogeny into a classification scheme may seem difficult at first sight. However, simple graphic methods may serve as powerful tools of analysis. Instead of working with dummy data, let us consider a simple case provided by Hinton (1966b). This example refers to the spiracles recorded in larvae and pupae of Psephenidae, a small family of Coleoptera. The larvae live in streams while the pupae are found either under water or on land close to the edges of streams. As a result of environmental instability, these larvae and pupae have evolved a wider variety of respiratory adaptations than any other family of beetles. The number of pairs of functional spiracles were recorded in different species by Hinton (1966b). He made a distinction between the early larval stases, which are quite similar in this respect, the last larval stase, and the pupa. Hinton's data are summarized in table 6.2, where, for convenience, taxa with the same respiratory system are clustered and are given an identification number. The number of evolutionary steps *sensu* Camin and Sokal (1965) and Gisin (1967) are given for each stase in the last column; practically, the number of evolutionary steps is equal to the number of spiracles which became nonfunctional.

Phyletic dendrograms may be constructed according to the methods proposed by Camin and Sokal (1965).[16] Dendrograms illustrated in figure 6.10 differ from cladograms in two points. As suggested by Gisin (1967) and put into practice by Gama (1971), the number of evolutionary steps is substituted for time on the vertical axis. This facilitates the passage into n-dimensional space described hereafter. The second concerns the common ancestry hypothesis underlying the construction of cladograms, a hypothesis developed in detail by Gaffney (1981). In the cladograms illustrated in figure 6.10, this hypothesis is replaced by the ancestor-descendant hypothesis as in another phyletic tree. This is related to the significance placed on vertices (on points). In cladograms, nodes represent synapomorphic resemblance (Nelson and Platnick 1981:146) while, in my dendrograms, they designate states of OTUs as indicated by the number of evolutionary steps. The dendrograms are in some respects kinds of morphoclines *sensu* Maslin (1952), whose polarity is determined. This view has some implications on the naming of taxa as outlined by Griffiths (1974). Figure 6.10E illustrates the cladogram equivalent to the dendrogram of figure 6.10D.

Dendrograms based on a single stase provide little information, whichever stase is selected (fig. 6.10 A-C). A dendrogram based on the entire ontogeny—i.e., on the total number of evolutionary steps found in each species—allows systematists to consider two major

Table 6.2. Spiracles in Larvae and Pupae of the Psephenidae

OTU	Stase	Thorax		Abdomen								Evolutionary steps in	
		II	III	1	2	3	4	5	6	7	8	stase	species
1	EL	o	—	—	—	—	—	—	—	—	o	8	19
	LL	o	—	—	—	—	—	—	—	—	o	8	
	P	—	—	o	o	o	o	o	o	o	—	3	
2	EL	o	—	—	—	—	—	—	—	—	o	8	25
	LL	o	—	—	—	—	—	—	—	—	o	8	
	P	—	—	—	—	—	—	—	—	o	—	9	
3	EL	—	—	—	—	—	—	—	—	—	o	9	22
	LL	—	—	—	—	—	—	—	—	—	o	9	
	P	—	—	—	o	o	o	o	o	o	—	4	
4	EL	—	—	—	—	—	—	—	—	—	—	10	23
	LL	—	—	—	—	—	—	—	—	—	o	9	
	P	—	—	—	o	o	o	o	o	o	—	4	
5	EL	—	—	—	—	—	—	—	—	—	—	10	24
	LL	—	—	—	—	—	—	—	—	—	—	10	
	P	—	—	—	o	o	o	o	o	o	—	4	

Source: From Hinton 1966
Open circles indicate functional spiracles.
EL: early larval stases; LL: last larval stase; P: pupa.

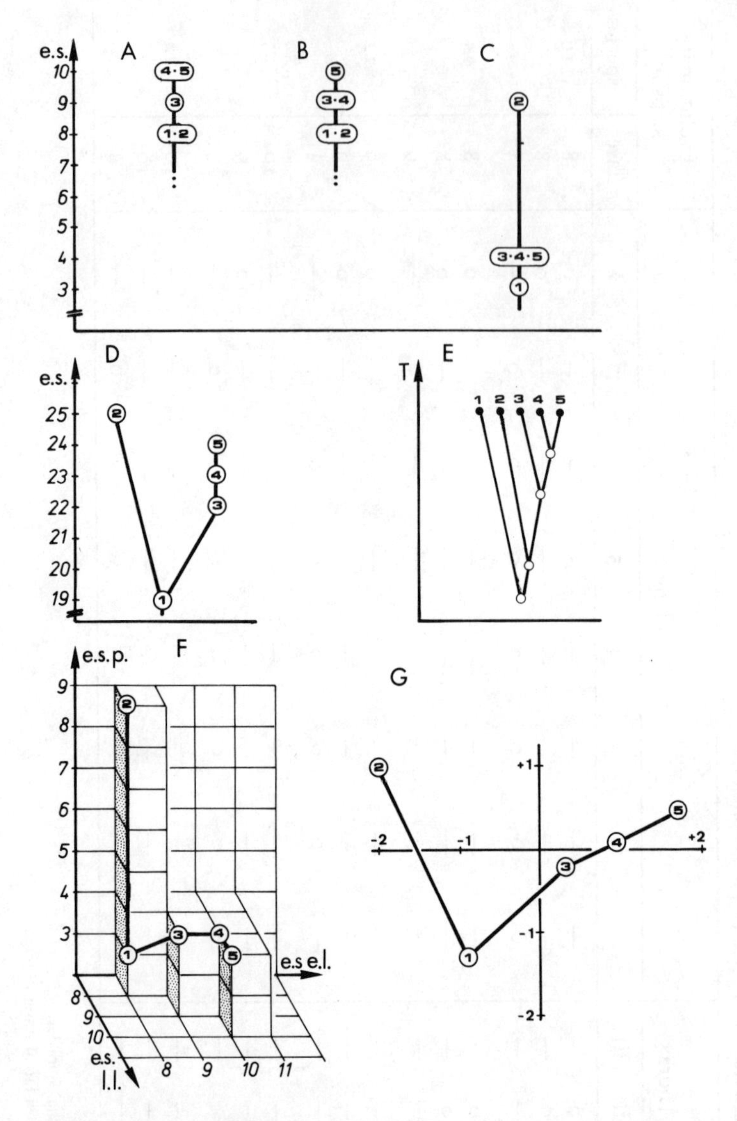

Figure 6.10. Ontogenetic dendrograms of some Psephenidae based on the presence of functional spiracles in the larvae and pupa (data from Hinton 1966b; see table 1). A: dendrogram based on the early larvae only; B: dendrogram for the last larva; C: dendrogram concerning the pupa; D: simple dendrogram based on the three stases; E: cladogram corresponding to dendrogram D; F: 3-dimensional ontogenetic dendrogram; G: result of PCA-ordination. OTUs are identified by the same number as in table 6.1 (e.s.: number of evolutionary steps; T: phylogenetic time; e.s.p.: e.s. for pupae; e.s.l.l.: e.s. for the last larvae; e.s.e.l.: e.s. for the early larvae).

trends from OTU 1 (fig. 6.10D). This dendrogram gives some information on the gap (number of evolutionary steps separating OTU 2 from OTU 1). However, no information is available on the distance between the terminal OTUs of the two branches (the angle between the branches is arbitrary), or on the ontogenetic level(s) responsible for the divergence. Therefore, a new dendogram is constructed in three dimensions, one per stase. Each stase for each species—more precisely its character state represented by the number of evolutionary steps—is plotted in this 3–dimensional diagram, to repeat stasis space (fig. 6.10F). The diagram, called an ontogenetic dendrogram, again reveals the existence of two trends. In addition it clearly demonstrates that the two branches are markedly diverging and that the divergence is due to the special respiratory pattern found in the pupa of OTU 2, which is isolated along the vertical axis (pupal axis). Finally, the diagram provides some information about the gap separating OTU 2 from OTU 5; OTU 2 is more distant from OTU 5 as it is from OTU 1, which was not obvious in fig. 6.10D. This gap might be estimated through any distance measure, since points are plotted in a space with homogeneous axes (all axes represent a discrete scale where similar evolutionary steps corresponding to nonfunctional spiracles are plotted).

However, the same kind of analysis applied, for instance, to a suite with five active stases would require a 5–dimensional diagram. In such a case, ordination techniques may serve to project the *n*-dimensional cladogram onto a space of two or three dimensions. Efficient projection of points in a space into fewer dimensions is the function of principal component analysis (PCA). As an example, Hinton's data were submitted to a PCA ordination (fig. 6.10G). OTU 2 is isolated from other OTUs along axis 1 which also opposes the pupal stase to larval stases. This clearly means that OTU 2 forms a divergent branch because of its particular pupa. The method thus allows the recognition of phyla and evolutionary trends, the identification of stases responsible for the divergences, and an estimation of the divergence magnitude.

A last comment concerns the exact meaning of links between two OTUs. These links mean only that one OTU is derived from the other. Particularly, it does not necessarily describe the evolutionary pathway followed between two species. This is illustrated in figure 6.10F where the link between OTUs 1 and 3 goes "across the corner" (the gap then is measured by an Euclidian metric) instead of going "round the edges" (as with a Manhattan metric). This special case corresponds to two situations. First, the data set is incomplete, as are Hinton's data. The intermediate step between OTUs 1 and 3 is unknown, and consequently these are three possible ways for the evolution to go from OTUs 1 to 3 "round the edges." As no information is given to choose between the three pathways, there is a direct link joining OTU 1 to OTU 3. In the second situation, the data set is complete but some "intermediate" steps are missing. The interpretation associated with this absence is that two or more characters are correlated, a possibility usually neglected by parsimony methods (Felsenstein 1982). In such a case, a Euclidian metric should be preferred to the Manhattan metric if the distance must be estimated. In any case, it must be noted that the PCA ordination used in figure 6.10 "optimizes" Euclidian distances.

One drawback of the ontogenetic dendrogram is that only one character is used in the analysis. In Hinton's data, this character is the number of functional spiracles, but it would not be possible to take into account any other characters with this approach.

Another drawback of the method is that two stases belonging to two species could be characterized as identical for instance, by nine evolutionary steps (i.e., the presence of only one pair of functional spiracles) but it might be that each species possesses a different pair of spiracles. In such a case, the result is biased. This point should be verified before applying the method.

Ontogenetic trajectories and systematics

An approach liable to supplement the ontogenetic dendrogram technique is based on the ontogenetic trajectories, described in the last section of the second part. The method also consists of plotting points in an *n*-dimensional space, but here points designate semaphoronts or stases. Thus, instead of working with OTUs as previously, an operational semaphoront is plotted in a character space with *n* dimensions, one dimension per character. Consequently, there is no limit on the number of characters which can be considered. Points representing the successive stases of a species are then joined to one another and the resulting vector—the ontogenetic trajectory—is a representation of the ontogeny.[17] Ontogenetic trajectories of different tydeid species belonging to subfamilies

Pronematinae and Triophtydeinae are illustrated in figure 6.11A; each mobile stase of each species is represented by a point which is characterized by the number of pairs of genital, aggenital, and eugenital setae. Each stase is thus plotted in a 3–dimensional space. The ontogenetic trajectories enable the recognition of two major trends corresponding to the two subfamilies. The diagram also shows that the different stases do not evolve at the same speed—at least for the characters taken into consideration. For instance, adults of Pronematinae have no eugenitals and thus have the same genital chaetotaxy as tritonymphs; this explains why, in figure 6.11, the ontogenetic trajectories of Pronematinae seem to have one step less than those of Triophtydeinae (the same point represents both the tritonymph and adult). The phylogenetic relations are suggested by dotted arrows and require a 3–dimensional version of the scheme published by Garstang (1922) and reproduced by de Beer (1958) to represent the relationships between ontogeny and phylogeny (fig. 6.11B). The idea that any Pronematinae adult is derived from another adult must be stressed. Indeed, it would be tempting to say that the *Homeopronematus* male, which has only one pair of aggenital seta, might be related to the protonymph found in most Tydeidae.[18] This is merely a case of retardation in the male (the female has four pairs of aggenitals), a condition that could lead to metathetely if the male happened to lose this single aggenital seta.

The meaning of the successive links of the ontogenetic trajectory must be summarized again. A link merely means that a stase succeeds another and is not intended to describe an evolutionary pathway through ontogeny with a Manhattan metric. As with ontogenetic dendrograms, gaps may separate stases (fig. 6.11). Gaps and missing stases can be explained either by an incomplete set of data or by a correlation between characters, as developed in the previous section. In the case of Tydeidae, the first explanation cannot be rejected totally even if a great deal of data are now available. But the second explanation cannot be rejected either, as it is well known that chaetotactic characters in mites are correlated (priority law, eustasy, etc.) and thus are not independent. Tydeidae are certainly not an exception (cf. André 1981a, b). This implies that a Manhattan metric is not necessarily the "best" one for measuring ontogenetic trajectories and that parsimony methods based on the independence of characters have to be used with care.

If more than three characters are considered, points are plotted into a

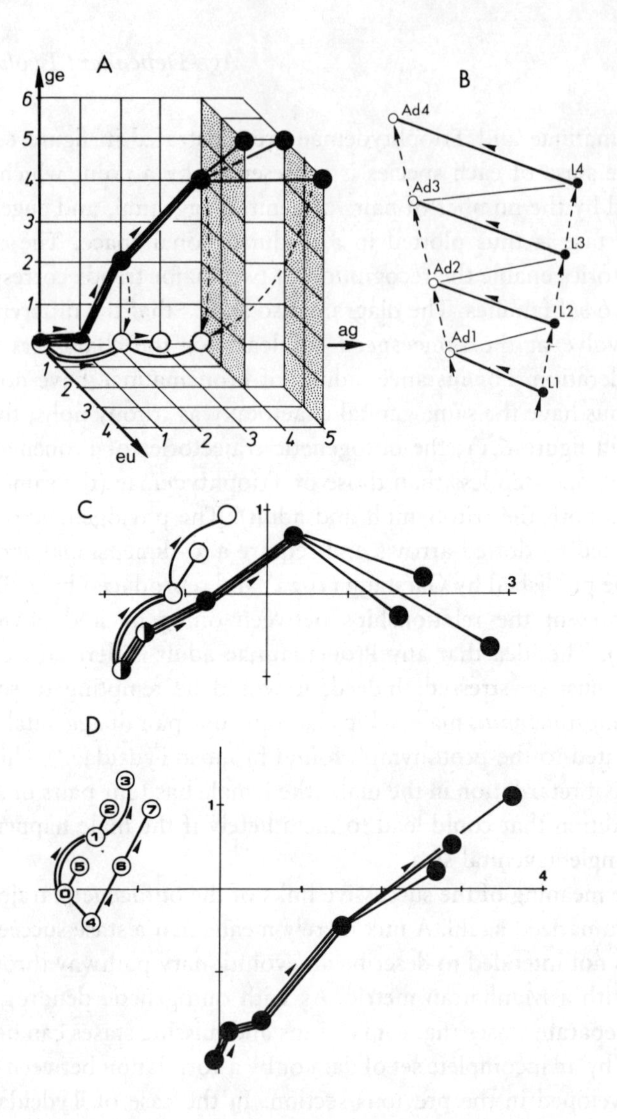

Figure 6.11. Ontogenetic trajectories (A,C,D) of Triophtydeinae (close circle) and Pronematinae (open circle) (Tydeidae). A: 3-dimensional diagram with trajectories based on eugenital (eu), aggenital (ag) and genital (ge) setae. C: PCA-ordination applied to data of diagram A. D: PCA-ordination considering, in addition to the 3 characters used in A and C, the number of eyes and setae on apotele I (number sequences discussed in the text). B: the relations between ontogeny and phylogeny (modified from Garstang, 1922); L1–L4: the succession of larvae; Ad1–Ad4: the succession of adults (phylogeny); L1–Ad1, L2–Ad2, the successions of ontogenies (same graphic conventions as in A: continuous lines refer to ontogeny; dashed lines, to phylogeny).

character space with *n*-dimensions, and it is again necessary to resort to PCA-ordination to project points into fewer dimensions. The PCA-ordination for figure 6.11A is shown in figure 6.11C. The linear sequence observed in the Pronematinae development is respected, but it is obvious that the two subfamilies are not clearly discriminated from each other and are diverging from a common origin (the larval level) if only the three characters cited previously are used. If the number of setae on apotele *I* (i.e., the claws and empodium) and the number of eyes are also taken into account, PCA-ordination yields the ontogenetic trajectories in figure 6.11D. Two sets of parallel trajectories are clearly distinct and correspond to the two subfamilies. Once again, the distortion of the linear ontogeny in Pronematinae is minimal. Two trends are also visible in Pronematinae, one corresponding to Pronematinae keeping an apotele *I* with an empodium (OSUs sequence 4–6–7) and the other without an apotele *I* (sequence 0–3). Note also that the figure illustrates an acceleration phenomenon. Indeed, the link (4–0) joins the two trends, and indicates that the larva of *Pronecupulatus* retains the empodium, while the older stases have lost it.

The ontogenetic trajectory method thus enables systematists not only to discriminate different taxa on the basis of their ontogeny, but also to highlight different peculiarities on their ontogeny. As in the previous method, it is based on heterochrony and discrete and discontinuous characters. The method is thus related to the approach proposed by Camin and Sokal (1965) and Gisin (1967), and is clearly associated with the stase concept presented in the first part. However, the resulting diagram may be more or less complex, depending on the number of characters and OSUs.

This method is merely my first attempt to incorporate ontogenetic data into systematics. Obviously, other methods of multivariate analysis, such as discriminant analysis, could be used with this approach, enabling biologists to gain a better understanding of evolutionary processes.

CONCLUSIONS

In their work on phylogenetic patterns and the evolutionary process, Eldredge and Cracraft (1980) were surprised that so little attention had been paid to ontogenetic data in twentieth-century systematic writings. Their opinion is confirmed by Hennig (1981) who considered that, in entomology, very little is known about the insect stages and their life histories. The same theme is heard in acarology when Balogy and Mahunka (1983) concluded that one of the most important tasks for future classification will be the detection and description of unknown developmental stages.

Obviously, the need for an ontogenetic approach in systematics is felt by biologists whether they are theoreticians or practicing systematists restricted to a well-defined taxon. At first sight, two distinct problems arise: one relates to the scarcity of data, the other to the distressing lack of a proper methodology allowing the treatment of ontogenetic data. In actual fact, both problems are two faces of the same coin. As emphasized by Kuhn (1970), *no natural history can be interpreted in the absence of at least some implicit body of intertwined theoretical and methodological belief that permits selection, evaluation, and criticism.* In the absence of such a paradigm, all the facts that could possibly pertain to the development of a given science are likely to seem equally relevant. As a result, fact-gathering is a near random activity and is usually restricted to the wealth of data that is readily at hand. Finally, this sort of fact-collecting produces a morass. This description by Kuhn is not unlike the situation that prevails today whenever the importance of postembryonic development in systematics is alluded to. If ontogenetic data are scarce or not always relevant, this is undoubtedly due to the lack of a theory, and its subsequent methodology, sufficient to guide research and data collecting, and not the converse.

My aim in writing this chapter was to provide biologists with a brief introduction to a paradigm likely to shed a new light on ontogenetic processes in arthropods. According to my approach two original techniques have been proposed. Other techniques, which might be contradictory, may be conceived in the future and other more sophisticated techniques will have to be developed. This is essential if biologists want to incorporate ontogeny into systematics and consider systematics as the study of living, and thus aging, beings rather than a collection of postage stamps. To conclude, Danser (1950) wrote rather directly on the subject: "The daily work of the systematist, 'systematic research' does not so much consist in the collecting of objects with a view to classifying them as in obtaining as complete a knowledge as possible of the life-cycle of the living beings which are to be compared later. The former is the primitive form of the latter."

ACKNOWLEDGMENTS

The major ideas underlying this article were read at the Third International Congress of Systematic and Evolutionary Biology held in Brighton from July 4–10, 1985. Some points were also presented at a seminar organized during the 1985 European Course in Acarology held at the University of Nottingham School of Agriculture. I wish to thank the staff of the course, especially G. O. Evans, for helpful discussion. I am grateful to Ph. Lebrun who provided much encouragement and help during the preparation of this paper. Early drafts of the manuscript were commented upon by C. J. Humphries, G. O. Evans and E. O. Wiley, and an anonymous referee who all provided many valuable criticisms. I wish also to thank C. J. Humphries for offering me the opportunity to publish this article, B. Hart for correcting the English and M. Beuckelaere for typing the successive drafts of the MS.

NOTES

1. Published in 1966.
2. This failure is fortunate because, as explained in the next section, the term "stase" had been coined previously in a quite different sense.
3. Actually, Hennig (1950) also distinguished a third category, called "Cyclomorphismus," which is not considered here.
4. There are few exceptions; for instance, Grassé (1949) wrote "nous appelerons *stade,* la période qui s'étend entre deux mues et *se caractérise par divers traits morphologiques*" (italics are mine).
5. Idionymy is the quality or the state of a particular organ as distinct from other organs of the same nature; consequently, such an organ can be named distinctly. However, Grandjean (1970:797) added in a footnote that "all or none" characters which ceased to be idionymic, but change in such a way that their mean number present discontinuous values, can theoretically suffice to discriminate stases—e.g., the facets of compound eyes.
6. Such a terminology, applied to insects, is discussed in detail by Snodgrass (1954), China et al. (1958), and Davies (1958). According to Andersson (1976), the terminology referring to the stases in Chilopoda is somewhat confusing; a synthesis is given in Andersson (1978). The one applied to crustaceans involves terms such as nauplius, metanauplius, etc.
7. The elattostasis is then termed "postpullus."

8. By proper ecdysis, I mean that a stase undergoes an ecdysis which is not concomitant with another.
9. The similarity between the two theories raises the question of whether Grandjean was aware of Garstang's theory. Grandjean owned a copy of *Embryologie et évolution*, the French translation by Rostand, of De Beer's *Embryology and Evolution*, in the margin of which he made numerous critical notes (Hammen 1981). Irrespective of the overall similarity between the two theories, Grandjean's final views represent an original approach likely to fill, at least in part, the gap between the study of ontogeny and the field of evolutionary genetics (Hammen 1981). Unfortunately, a detailed comparative study of the theories is beyond the scope of this paper.
10. Strictly speaking, the presence of vestigial claws is not an "all or none" character and the example could be reduced to a 2–state character (presence/absence of apotele *I*). However, this does not change the value of the example.
11. These papers are rather theoretical and give rather few examples of new evidence except for Alberch (1985). Most examples are given to support the views expressed by the authors, and at best refer to examples cited by de Beer. Furthermore, it must be noted that the debate is between biologists involved in the study of vertebrates and little attention has been paid to arthropods.
12. The hologeny or total development of organisms is conceived as a series of ontogenetic spirals developing through time.
13. The definition given by Matsuda (1976:22) conforms with de Beer's definition.
14. Gould (1977:304) noted that this classification duplicates the distinction between progenesis (prothetely) and neoteny (metathetely). However, the terms mentioned above are peculiar to animals with stases. Gould (1977) also noticed that this distinction is not accepted by all entomologists, e.g. Matsuda (1979:139), who related prothetely to neoteny.
15. Zeugloptera are lowered in hierarchic rank and considered to be a suborder of Lepidoptera by Kristensen (1971, 1975). Kristensen's arguments are based solely on morphological characters and do not diminish the value of the example.
16. Other parsimony methods could be used. The two premises adopted here conform with the stase approach and are (1) the character state polarity is known and (2) reversion from a derived state to the ancestral state is impossible. These assumptions are not necessarily accepted by all authors and are not compatible with other methods such as Wagner-analysis (see Felsenstein 1982).
17. My concept of ontogenetic trajectory is related to species development and is thus essentially different from that of Alberch et al. (1979), which is an idealization referring to the growth of a simple individual in a population.

18. Indeed, the most frequent genital formula in a tydeid protonymph involves only a single pair of aggenital setae (André 1981a).

REFERENCES

Alberch, P. 1985. Problems with the interpretation of developmental sequences. *Syst. Zool.* 34: 46–58.

Alberch, P., S. J. Gould, G. F. Oster, and D. B. Wake. 1979. Size and shape in ontogeny and phylogeny. *Paleobiology,* 5:296–317.

Anderson, W. H., G. Anastos, R. W. Bunn, B. F. Eldridge, R. K. Latta, and M. R. Wheeler. 1971. Stadium, stage, and instar. *Bull. ent. Soc. Am.* 17:17.

Andersson, G. 1976. Post-embryonic development of *Lithobius forficatus* (Chilopoda: Lithobiidae). *Ent. scand.* 7: 161–168.

Andersson, G. 1978. An investigation of the post-embryonic development of the Lithobbidae: some introductory aspects. *Abh. Verh. naturw. Ver. Hamburg* (NF), 21/22: 63–71.

André, H. M. 1979. A generic revision of the family Tydeidae (Acari: Actinedida). I. Introduction, paradigms and general classification. *Annls Soc. r. Zool. Belg.* 108:189–208.

André, H. M. 1981a. A generic revision of the family Tydeidae (Acari: Actinedida). II. Organotaxy of the idiosoma and gnathosoma. *Acarologia,* 22:31–46.

André, H. M. 1981b. A generic revision of the family Tydeidae (Acari: Actinedida). III. Organotaxy of the legs. *Acarologia,* 22:165–178.

André, H. M. 1987. The concept of stase in Collembola. *Proceedings VI. intl Congress on Apterygota, Moscow (1985):* (in press).

André, H. M. and R. Jocqué. 1987. The definition of stases in spiders and other arachnids. 1985. In L. Baerts, R. Jocqué and J.P. Maefait, eds. IXème Colloque européen d'Arachnologie. *Mém. Soc. r. Belge Ent.* 33:1–14.

André, H. M., Ph. Lebrun, and S. Leroy. 1984. The systematic status and geographical distribution of *Camisia segnis* (Acari: Oribatida). *Internat. J. Acarol.* 10:153–158.

Audy, J. R., M. Nadchatram, and P. H. Vercammen-Grandjean. 1963. La "néosomie," un phénomène de néoformation en acarologie, alliée à un cas remarquable de tachygenèse. *Bull. Acad. roy. Belg. (Sci.)* 49:1015–1027.

Audy, J. R., F. J. Radovsky and P. H. Vercammen-Grandjean. 1972. Neosomy: radical instrastadial metamorphosis associated with arthropod symbioses. *J. Med. Ent.* 9:487–494.

Balogh, J. and S. Mahunka. 1983. *Primitive Oribatids of the Paleartic Region.* Amsterdam: Elsevier.

Betsch, J.-M. 1986. Discussion following the contribution by André (1987).

Boczek, J., C. Jura, and A. Krzysztofowicz. 1969. The comparison of the structure of the internal organs of postembryonic stages of *Acarus farris* (Oud.) with special reference to the hypopus. In G. O. Evans, ed., *Acarology, Proceedings 2nd Intl Congr. Acarol., Sutton Bonington, 1967*, pp. 265–271. Budapest: Akadémiai Kiadó

Bounhiol, J.-J. 1980. *Larves et métamorphoses.* Paris: Presses Universitaires de France.

Bourgeois, A. 1971. Le polymorphisme des Hypogastruridae et les phénomènes de "gynomorphose" dans ce groupe. *Rev. Ecol. Biol. Soc,* 8:139–144.

Bourgeois, A. 1974. Nouveaux cas d'épitoque chez les Collemboles Hypogastruridae. *Pedobiologia* 14:191–195.

Bourgeois, A. and P. Cassagnau. 1973. Les perturbations de type epitoque ches les Collemboles Hypogastruridae. *C. R. Acad. Sci. Paris,* 277D:1197–1200.

Brien, P. 1974. *Le vivant. Epigenèse. Evolution épigénétique.* Bruxelles: Editions de l'université de Bruxelles.

Brooks, D. R. and E. O. Wiley. 1985. Theories and methods in different approaches to phylogenetic systematics. *Cladistics,* 1:1–11.

Camin, J. H. and R. R. Sokal. 1965. A method for deducing branching sequences in phylogeny. *Evolution* 19:311–326.

Canard, A. 1984. *Contribution à la connaissance du développement, de l'écologie et de l'écophysiologie des Aranéides des landes armoricaines.* D. Sc. thesis, Université de Rennes.

Carlson, R. W. 1983. Instar, stadium, and stage: definitions to fit usage. *Ann. entomol. Soc. Am.* 76:319.

Carlsson, G. 1962. Studies on Scandinavian black flies. *Opusc. Ent. suppl.* 21:1–280.

Carothers, E. E. 1923. Notes on the taxonomy, development and life history of certain Acrididae (Orthoptera). *Trans. Am. ent. Soc.* 49:7–24.

Cassagnau, P. 1956. L'influence de la températuee sur l'apparition du "genre" *Spinisotoma* (Collembole, Isotomidae). *C.R. Acad. Sci. Paris,* 242D:1531–1534.

Cassagnau, P. 1972. Adaptation écologique et morphogenèse: les écomorphoses. *Atti. IX Congr. Naz. Ital. Ent., Bertelli, Firenze,* pp. 227–244.

Cassagne-Méjean, F. 1969. Sur les calyptostases des Hydrachnelles. In G. O. Evans, ed. *Acarology, Proceedings 2nd Intl Congr. Acarol., Sutton Bonington, 1967,* pp. 93–97. Budapest: Akademiai Kiado.

Chapman, R. F. 1971. *The Insects. Structure and Function.* 2nd ed.). London: English Universities Press.

China, W. E., B. Henson, B. M. Hobby, H. E. Hinton, T. T. Macan, O. W. Richards and V. V. Wigglesworth. 1958. The terms "larva" and "nymph" in entomology. *Trans. Soc. Br. Ent.* 13:17–24.

Cohen, J. and B. D. Massey. 1983. Larvae and the origins of major phyla. *J. Linn. Soc. Biol.* 19:321–328.

Coineau, Y. 1974. Eléments pour une monographie morphologique, écologique et biologiques des Caeculidae (Acariens). *Mém. Mus. natn. Hist. nat.* (N.S.) 81:1–299.

Coineau, Y. 1979. Les Adamystidae, une étonnante famille d'acariens prostigmates primitifs. In E. Piffl, ed., *Acarology, Proceedings of the 4th Intl Congr. Acarol., Saalfelden, 1974,* pp. 431–435. Budapest: Akadémiai Kiadó.

Danser, B. H. 1950. A theory of systematics. *Bibl. Biotheor.* 4:117–180.

Davies, R. G. 1958. The terminology of the juvenile phases of insects. *Trans. Soc. Br. Ent.* 13:25–36.

de Beer, G. R. 1930. *Embryology and evolution.* Oxford: Clarendon Press.

de Beer, G. R. 1958. *Embryos and ancestors* (3rd ed.). Oxford: Clarendon Press.

de Queiroz, K. 1985. The ontogenetic method for determining character polarity and its relevance to phylogenetic systematics. *Syst. Zool.* 34:280–299.

Eldredge, N. and J. Cracraft. 1980. *Phylogenetic Patterns and the Evolutionary Process.* New York: Columbia University Press.

Elsen, P., A. Fain, M.-C. Henry and M. DeBoeck. 1983. Description morphologique de divers membres du complexe *Simulium damnosum* provenant du Zaïre occidental. *Revue Zool. afr.* 97:653–673.

Emden, F. I. van, 1927. Was hat di entomologische Larvensystematik den anderen entomologischen Disziplinen gegeben, bezw. kann sie ihnen geben? *Ent. Mitt.* 16:12–16.

Emden, F. I. van, 1957. The taxonomic significance of the characters of immature insects. *Ann. Rev. Ent.* 2:91–106.

Evans, G. O., D. A. Griffiths, D. Macfarlane, P. W. Murphy, and W. M. Till. 1985. *The Acari. A Practical Manual.* vol. 1, Sutton Bonington: University of Nottingham School of Agriculture.

Evans, G. O. and W. M. Till. 1979. Mesostigmatid mites of Britain and Ireland (Chelicerata: Acari - Parasitiformes). An introduction to their external morphology and classification. *Trans. zool. Soc. Lond.* 35:139–270.

Fabre, H. 1857. Mémoire sur l'hypermétamorphose et les Moeurs des Méloides. *Annls Sci. Nat. Zool.* 7:299–365.

Fain, A. 1957. Sur la position systématique de *Ricardoella eweri* Lawr. et de *Boydaia angelae* Wom. Remaniement de la famille Ereynetidae. *Revue Zool. Bot. afr.* 55:249–252.

Fain, A. 1963. Chaetotaxie et classification des Speleognathinae. *Bull. Inst. R. Sci. Nat. Belg.* 39(9):1–80.

Fain, A. 1967. Les hypopes parasites des tissus cellulaires des oiseaux (Hypodectidae: Sarcoptiformes). *Bull. Inst. r. Sci. nat. Belg.* 43(4):1–139.

Fain, A. 1969. Adaptation to parasitism in mites. *Acarologia* 11:429–449.

Fain, A. 1972. Développement post-embryonnaire chez les acariens de la sous-famille Speleognathinae (Ereynetidae: Trombidifomes). *Acarologia,* 13:607–614.

Fain, A. and J. Bafort. 1967. Cycle évolutif et morphologie de *Hypodectes (Hypodectoides) propus* (Nitzsh) acarien nidicole à deutonymphe parasite tissulaire des pigeons. *Bull. Acad. Roy. Sci. Belg.* 53:501–533.

Felsenstein, J. 1982. Numerical methods for inferring evolutionary trees. *Q. Rev. Biol.* 57:379–404.
Fink, T. J. 1983. A further note on the use of the terms instar, stadium, and stage. *Ann. Entomol. Soc. Am.* 76:316–318.
Frankel, G. and G. Bhaskaran. 1973. Pupariation and pupation in Cyclorraphous flies (Diptera): terminology and interpretation. *Ann. Entomol. Soc. Am.* 66:418–422.
Gaffney, E. S. 1981. An introduction to the logic of phylogeny reconstruction. In J. Cracraft and N. Eldredge, eds., *Phylogenetic Analysis and Paleontology,* pp. 79–111. New York: Columbia University Press.
Gama, M. M. da, 1971. Application de la méthode de la "systématique idéale" à quelques espèces du genre *Xenylla. Rev. Ecol. Biol. Sol* 8:189–193.
Garstang, W. 1922. The theory of recapitulation: a critical restatement of the biogenetic law. *J. Linn. Soc. Zool.* 35:81–101.
Garstang, W. 1928. The morphology of the Tunicata and its bearing on the phylogeny of the Chordata. *Quart. J. Microsc. Sci.* 87:103–193.
Garstang, W. 1966. *Larval forms with other zoological verses.* Oxford: Blackwell.
Giard, A. 1905. La poecilogonie. *Bull. Sci. France Belg.* 39:153–187.
Girard, M. 1873. *Traité élémentaire d'entomologie.* vol. 1. Paris: Librairie Baillières and et fils.
Gisin, H. 1967. La systématique idéale. *Z. Zool. Syst. Evolforsch.* 5:111–128.
Gould, S. J. 1927. *Ontogeny and Phylogeny.* Cambridge: Harvard University Press.
Grandjean, F. 1938a. Sur l'ontogénie des acariens. *C. R. Acad. Sci. Paris,* 206D:146–150.
Grandjean, F. 1938b. Au sujet de la néoténie chez les acariens. *C. R. Acad. Sci. Paris,* 207D:1347–1351.
Grandjean, F. 1939. Observations sur les Oribates (12e série). *Bull. Mus. natn Hist. nat.* 11:300–307.
Grandjean, F. 1940. Observations sur les Oribates (15e série). *Bull. Mus. natn. Hist. nat.* 12:332–339.
Grandjean, F. 1942. Observations sur les Labidostommidae (3e série). *Bull. Mus. Natn. Hist. nat.* 14:319–326.
Grandjean, F. 1945. Observations sur les Acariens (8e série). *Bull. Mus. natn. Hist. nat.* 17:399–406.
Grandjean, F. 1947a. L'harmonie et la dysharmonie chronologiques dans l'évolution des stases. *C. R. Acad. Sci. Paris,* 225D:1057–1060.
Grandjean, F. 1947b. Au sujet des Erythroides. *Bull. Mus. natn. Hist. nat.* 19:327–334.
Grandjean, F. 1951a. Comparison du genre *Limnozetes* au genre *Hydrozetes* (Oribates). *Bull. Mus. natn. Hist. nat.* 23:200–207.
Grandjean, F. 1951b. Les relations chronologiques entre ontogenèses et phylogenèses d'après les petits caractères discontinus des Acariens. *Bull. biol. France Belg* 85:269–292

Grandjean, F. 1954a. Observations sur les Oribates (31e série). *Bull. Mus. natn. Hist. nat.* 26:582–589.
Grandjean, F. 1954b. Les deux sortes de temps et l'évolution. *Bull. Biol. France Belg.* 88:413–434.
Grandjean, F. 1957a. Les stases du développement ontogénétique chez *Balaustium florale* (Acarien, Erythroide). Première partie. *Annls. Soc. ent. France* 125:135–152.
Grandjean, F. 1957b. L'évolution selon l'âge. *Archs Sci. Genève,* 10:477–526.
Grandjean, F. 1958. Sur le comportement et la notation des poils accessoires postérieurs aux tarses des Nothroides et d'autres Acariens. *Archs Zool. Exp. Gen.* 96:277–308.
Grandjean, F. 1962. Prélarves d'Oribates. *Acarologia,* 4:423–439.
Grandjean, F. 1970. Stases. Actinopiline. Rappel de ma classification des Acariens en trois groupes majeurs. Terminologie en soma. *Acarologia* 11:796–827.
Grassé, P.-P. 1949. Ordre des Isoptères ou Termites. In P.-P. Grassé, ed., *Traité de Zoologie* vol. 10. pp. 408–544. Paris: Masson.
Grégoire-Wibo, C. 1974. Bioécologie de *Folsomia quadrioculata* (Insecta, Collembola). *Pedobiologia* 14:199–207
Griffiths, G. C. D. 1974. On the foundations of biological systematics. *Acta Biotheor.* 23:85–131.
Hammen, L. van der, 1964. The relation between phylogeny and post-embryonic ontogeny in actinotrichid mites. *Acarologia* 6 (f.h.s.):85–90.
Hammen, L. van der, 1974. L'évolution du cycle vital chez les Acariens et les autres groupes d'Arachnides. *Acarologia* 15:384–390.
Hammen, L. van der, 1975. L'évolution des Acariens et les modèles de l'évolution des Arachnides. *Acarologia* 16:377–381.
Hammen, L. van der, 1978. The evolution of chelicerate life-cycle. *Acta Biotheor.* 27:44–60.
Hammen, L. van der, 1979. Evolution in mites, and the patterns of evolution in Arachnidea. In E. Piffl, ed., *Acarology, Proceedings 4th Intl Congr. Acarol., Saalfelden,* 1974, pp. 425–430. Budapest: Akadémiai Kiadó.
Hammen, L. van der, 1980. *Glossary of acarological terms.* Vol. 1: *General Terminology.* The Hague: Junk.
Hammen, L. van der, 1981. Numerical changes and evolution in actinotrichid mites (Chelicerata). *Zool. Verh. Leiden* 182:1–47.
Henking, H. 1882. Beiträge zur Anatomie, Entwicklungsgeschichte und Biologie von *Tromibidium fuliginosum. Z. Wiss. Zool.* 37:553–663.
Hennig, W. 1950. *Grundzüge einer Theorie der Phylogenetischen Systematik.* Berlin: Deutscher Zentralverlag.
Hennig, W. 1966. *Phylogenetic Systematics.* Urbana: University of Illinois Press.
Hennig, W. 1969. *Die Stammesgeschichte der Insekten.* Frankfurt am Main: Waldemar.
Hennig, W. 1981. *Insect Phylogeny.* Chichester: John Wiley and Sons.

Hinton, H. E. 1946a. Concealed phases in the metamorphosis of insects. *Nature (Lond.)* 157:552–553.
Hinton, H. E. 1946b. On the homology and nomenclature of the setae of lepidopterous larvae, with some notes on the phylogeny of Lepidoptera. *Trans. R. Ent. Soc. Lond.* 97:1–37.
Hinton, H. E. 1948. On the origin and function of the pupal stage. *Trans. R. ent. Soc. Lond.* 99: 395–409.
Hinton, H. E. 1958. Concealed phases in the metamorphosis of insects. *Sci. Progr. (Lond.)* 46: 260–275.
Hinton, H. E. 1966a. Apolysis in arthropod moulting cycles. *Nature (Lond.)* 211: 871.
Hinton, H. E. 1966b. Respiratory adaptations of the pupae of beetles of the family Psephenidae. *Phil. Trans. R. Soc.* ser. B, 251: 211–245.
Hinton, H. E. 1971. Some neglected phases in metamorphosis. *Proc. R. Ent. Soc. Lond.* (C), 35: 55–64.
Hinton, H. E. 1973. Neglected phases in metamorphosis: a reply to V. B. Wigglesworth. *J. Ent.* (A), 48: 57–68.
Hinton, H. E. 1976. Notes on neglected phases in metamorphosis, and a reply to J. M. Whitten. *Ann. entomol. Soc. Am.* 69: 560–566.
Hinton, H. E. an I. M. Mackerras. 1970. Reproduction and metamorphosis. In *The Insects of Australia,* pp. 83–106. Carlton, Victoria: Melbourne University Press.
Imms, A. D. 1948. *A general textbook of entomology* (7th ed.) London: Methuen.
Istock, C. A. 1967. The evolution of complex life cycle phenomena: an ecological perspective. *Evolution* 21: 592–605.
Jacob, F. 1970. *La logique du vivant.* Paris: Gallimard.
Jägersten, G. 1972. *Evolution of the Metazoan Life Cycle.* London: Academic Press.
Joly, P. 1977. Le développement postembryonnaire des insectes. In P.-P. Grassé, ed., *Traité de Zoologie,* vol. 8, pp. 409–657. Paris: Masson.
Jones, J. C. 1978. A note on the use of the terms instar and stage. *Ann. entomol. Soc. Am.* 71: 491–492.
Jones, J. C. 1983. A note on the life history of insects. *Ann. Entomol. Soc. Am.* 76: 320–321.
Juberthie, C. 1955. Sur la croissance post-embryonnaire des Aranéides: croissance linéaire du corps et dysharmonie de croissance des appendices. *Bull. Soc. Hist. nat. Toulouse* 90: 83–102.
Kluge, A. G. 1985. Ontogeny and phylogenetic systematics. *Cladistics* 1: 13–27.
Koestler, A. 1972. The tree and the candle. In W. Gray and N. D. Rizzo, eds., *Unity Through diversity.* New York: Gordon 1 Breach Science Publ.
Koestler, A. 1967. *The Ghost in the Machine.* London: Macmillan.
Kollman, J. 1885. Das Ueberwintern von europaïschen Froschund Triton larven und die Umwandlung des mexicanischen Axolotl. *Verh. Naturf. Ges. Basel* 7: 387–398.

Krantz, G. W. 1978. *A Manual of Acarology*, 2nd ed., Corvallis: Oregon State University Book Stores.

Kuhn, T. S. 1970. *The structure of scientific revolutions*. 2nd ed. Chicago: University of Chicago Press.

Kunckel d'Herculaïs, J. 1890. Mécanismes physiologiques de l'éclosion des mues et de la métamorphose chez les insectes orthoptères de la famille des acridides. *C. R. Acad. Sci. Paris* 110D: 807–809.

Kristensen, N. P. 1971. The systematic position of the Zeugloptera in the light of recent anatomical investigations. *Proc. 13th Congr. Ent. Moscow 1968*. 1: 261. Leningrad: Publ. House Nauka.

Kristensen, N. P. 1975. The phylogeny of hexapod "orders." A critical review of recent accounts. *Z. Zool. Syst. Evolut.-forsch.* 13: 1–44.

Kristensen, N. P. 1981. Phylogeny of insect orders. *Ann. Rev. Entomol.* 26: 135–157.

La Baume, W. 1918. Biologie der marokkanischen Wander heuschrecke (*Stauronotus maroccanus* Thunb.) Beoachtungen aus Kleinasien in der Hahren 1916 u. 1917. *Monogr. Angew. Ent.* 3: 157–274.

Lacordaire, Th. 1834. *Introduction à l'entomologie*. vol. 1. Librairie encyclopedique de Roret.

Lenz, F. 1926. Die Chironomiden-Metamorphose in ihrer Bedeutung tür di Systematik. *Ent. Mitt.* 15: 440–442.

Leppick, E. E. 1970. Phylogeny, hologeny and coenogeny, basic concepts of environmental biology. *Acta Biotheor.* 23: 170–193.

Lewis, T. 1973. *Thrips: Their Biology, Ecology, and Economic Importance*. London: Academic Press.

Lindquist, E. E. 1984. Current theories on the evolution of major groups of Acari and on their relationships with other groups of Arachnida, with consequent implications for their classification. In D. A. Griffiths and C. E. Bowman, eds., *Acarology VI* 1: 28–62. Chichester: Ellis Horwood.

Løvtrup, S. 1974. On von Baerian and Haeckelian recapitulation. *Syst. Zool.* 27: 348–352.

Lubbock, J. 1873. *On the Origin and Metamorphoses of Insects*. London: Macmillan.

Maslin, T. P. 1952. Morphological criteria of phyletic relationships. *Syst. Zool.* 1: 49–70.

Matsuda, R. 1976. *Morphology and Evolution of Insect Abdomen*. Oxford: Pergamon Press.

Matsuda, R. 1979. Abnormal metamorphosis and arthropod evolution. In A. P. Gupta, ed., *Arthropod phylogeny*, pp. 137–256. New York: Van Nostrand.

Meredith, S. O. E. 1984. Methods of identification of members of the *Simulium damnosum* complex in West Africa and problems associated therewith. *Report 20th meeting of the scientific working group on filiariasis, W. H. O.*: 16.

Nelson, G. 1978. Ontogeny, phylogeny, paleontology and the biogenetic law. *Syst. Zool.* 27: 324–345.

Nelson, G. 1985. Outgroups and ontogeny. *Cladistics* 1: 29–45.

Nelson, G. and N. Platnick. 1981. *Systematics and Biogeography; Cladistics and Vicariance.* New York: Columbia University Press.

Newell, I. M. 1973. The protonymph of *Pimeliaphilus* (Pterygosomatidae) and its significance to the calyptostases in the Parasitengona. In M. Daniel and B. Rosicky, eds., *Proceedings 3rd Intl Congr. Acarol., Prague, 1971*, pp. 789–795. The Hague: Junk.

Newport, G. 1947. On the natural history, anatomy and development of the oil beetle, *Meloë*, more especially of *Meloë cicatricosus* Leach. *Trans. Linn. Soc. Lond.* 20: 297–355.

Okada, M. 1958. Embryonic moulting in the rice stem borer *Chilo suppressalis. Jap. J. Appl. Ent. Zool.* 2: 295–296.

Pierce, W. D. 1964. The Strepsiptera are a true order, unrelated to Coleoptera. *Ann. entomol. Soc. Am.* 57: 603–605.

Roonwal, M. L. 1952. Further observations on directional changes in locusts and other short-horned grasshoppers (Insecta: Orthoptera: Acrididae), and the importance of the third instar. *Proc. Nat. Inst. Sci. India* 18: 217–232.

Rambla, M. 1980. Neoteny in Opilions. In J. Gruber, ed., *8. Internationaler Arachnologen-Kongress, Verhandlungen, Wien, 1980*, pp. 489–492. Wien: Verlag Egermann.

Schaefer, C. W. 1983. Instar, stadium, and stage: A new look at old questions. *Ann. entomol. Soc. Am.* 76: 315.

Selander, R. B. and J. M. Mathieu. 1964. The ontogeny of blister beetles (Coleoptera Meloidae). I. A study of three species of the genus *Pyrota. Ann. entomol. Soc. Am.* 57: 711–732.

Selander, R. B. and R. C. Weddle. 1969. The ontogeny of blister beetles (Coleoptera Meloidae). II. The effects of age of triungulin larvae at feeding and temperature on development in *Epicauta segmenta. Ann. Entomol. Soc. Am.* 62: 27–39.

Simon, H. J. 1965. The architecture of complexity. *Gen. Systems* 10: 63–76.

Sing-Pruthi, H. 1924. Studies on insect metamorphosis. 1. Prothetely in mealworms *(Tenebrio molitor)* and other insects: Effects of different temperatures. *Biol. Rev.* 1: 139–147.

Sitnikova, L. G. 1960. Les prélarves des Oribates (in Russian, French summary). *Parazit. Sb.* 19: 220–236.

Snodgrass, R. E. 1935. *Principles of Insect Morphology.* New York: McGraw Hill.

Snodgrass, R. E. 1954. Insect metamorphosis. *Smithson. Misc. Collns.* 122(9): 1–124.

Steyskal, G. C. 1984. A linguistic look at the stage-stadium-instar problem. *Ann. Entomol. Soc. Am.* 77: iii.

Thom, R. 1977. *Stabilité structurelle et morphogenèse.* 2d ed. Paris: InterEditions.

Travé, J. 1964. Importance des stases immatures des Oribates en systématique et en écologie. *Acarologia,* 6 (f.h.s.): 47–54.

Uvarov, B. P. 1928. *Locusts and Grasshoppers*. London: Imperial Bureau of Entomology.
Uvarov, B. 1966. *Grasshoppers and Locusts* (vol. 1). Cambridge: Cambridge University Press.
Vachon, M. 1934. Sur le développement postembryonnaire des Pseudoscorpionides. Première note: les formes immatures du *Chelifer cancroides* L. *Bull. Soc. Zool. Fr.* 59: 134–160.
Vachon, M. 1953. Commentaires à propos de la distinction des stades et des phases de développement postembryonnaires chez les araignées. *Bull. Mus. Natn. Hist. Nat.* 25: 294–297.
Van Impe, G. 1985. *Contribution à la conception de stratégies de contrôle de l'acarien tisserand commun, Tetranychus urticae Koch (Acari: Tetranychidae)*. D. Sc. thesis. Université Catholique de Louvain.
Vercammen-Grandjean, P. H. 1969a. Repository of phylogeny and taxonomy in acarines. In G. O. Evans, ed., *Acarology, Proceedings 2nd Intl Congr. Acarol., Sutton Bonington, 1967*, pp. 107–112. Budapest: Akademiai Kiado.V
Vercamen-Grandjean, P. H. 1969b. Le stade larvaire, sanctuaire de la phylogénie et de la taxinomie ches les Acariens. *Annls Paras. Hum. Comp.* 44: 205–210.
Vercammen-Grandjean, P. H., R. L. Langston and J. R. Audy. 1973. Tentative nepophylogeny of trombiculids. *Folia Parasit.* 20: 49–66.
Verhoeff, C. 1905. Gliederfüssler: Arthropoda. IV. Entwicklung. *Bronn's Kl. Ordn Tierreichs*, 5: 115–215.
Verhoeff, K. W. 1919. Zur Entwicklung Morphologie und Biologie der Vorlarven and Larven der Canthariden. *Arch. Naturg.* (A), 83: 102–140.
Voorzanger, B. and W. J. van der Steen. 1982. New perspectives on the biogenetic law? *Syst. Zool.* 31: 202–205.
Walter, C. 1920. Die Bedeutung der Apodermata in der Epimorphose der Hydracarina. *Festsch. der 60. Beburtstages von F. Zschokke, Basel*, 24:
Weber, H. 1933. *Lehrbuch der Entomologie*. Jena: Fisher.
Whitten, J. M. 1976. Definition of insect instars in terms of "apolysis" or "ecdysis." *Ann. Entomol. Soc. Am.* 69: 556–559.
Wigglesworth, V. B. 1954. *The Physiology of Insect Metamorphosis*. Cambridge: Cambridge University Press.
Wigglesworth, V. B. 1973. The significance of "apolysis" in the moulting of insects. *J. Ent.* (A), 47: 141–149.
Wilbur, H. M. 1980. Complex life cycles. *Ann. Rev. Ecol. Syst.* 11: 67–93.
Wiley, E. O. 1981. *Phylogenetics: The Theory and Practice of Phylogenetic Systematics*. New York: Wiley-Interscience.
Witte, H. 1978. Die postembryonale Entwicklung und die funktionelle Anatomie des Gnathosoma in der Milbenfamilie Erythraeidae (Acarina, Prostigmata). *Zoomorphologie* 91: 157–189.
Zimmerman, W. 1948. *Grundfragen der Evolution*. Frankfurt am Main.

7. Epigenetics

Søren Løvtrup

What are the connections between "ontogeny", "systematics" and "epigenetics"? If we look in Collins English Dictionary, we find that "ontogeny" is "the entire sequence of events involved in the development of an individual organism," "systematics" is the "study of systems and the principles of classification and nomenclature," and "epigenesis" ("epigenetics" is not to be found) is "the widely accepted theory that an individual animal or plant develops by the gradual differentiation and elaboration of a fertilized egg cell."

The last definition refers to the classical, but unfounded (Løvtrup 1987), antithesis between preformation and epigenesis. Nevertheless, in the present context it is quite acceptable because it allows for the definition of "epigenetics" as "the study of the mechanisms responsible for the effectuation of ontogenetic development." This definition is, I believe, in close agreement with the intentions of Waddington when he originally coined the expression.

It is evident, however, that it will be impossible to establish any connection with either "ontogeny" or "epigenetics" and "systematics," and therefore we must return to the dictionary where we find that "taxonomy" is "the branch of biology concerned with the classification of organisms into groups." In general, biologists do not observe the distinction between" "systematics" and "taxonomy," and yet it is evident that only the latter concept is of interest in our case.

In his *Philosophie Zoologique* (1809) Lamarck proposed to discriminate between "classification" and "distribution générale"; the former concept refers to traditional Linnean taxonomy—or systematics—and the latter to "l'ordre . . . de la nature"—the course of evolution. The reality of evolution has been accepted almost unanimously since the publication of Darwin's *On the Origin of Species* in 1859. Yet taxonomy continued along the tracks staked out by the Linnean conventions for more than a century before biologists began to seriously consider the consequences of the distinction pointed out by Lamarck.

This time again it was a book that stemmed the tide—Willi Hennig's *Phylogenetic Systematics* (1966), an English translation and revision of a work previously published in German (1950). Hennig stated certain consequences imposed upon classification by the theory of evolution, thereby founding the discipline of phylogenetic taxonomy. Unquestionably, the most unexpected result of this contribution is that phylogenetic hierarchies consist of dendrograms which to a large extent are dichotomous. Claiming to represent the historical course of evolution, these dendrograms evidently must represent sequences of events involved in the evolution of the living world, each bifurcation representing the origin of two separate phylogenetic lineages from a common ancestral form.

Phylogenetic classifications are incomplete histories of evolution. This incompleteness has two aspects, only the first of which is remediable. The first depends on the fact that many living beings have not yet been classified phylogenetically, and the second on the fact that the history of life in the past will forever be known only fragmentarily. However, if we disregard these shortcomings it is seen that ontogeny and phylogenetic classification are both concerned with sequences of events taking place in living organisms. Is there any relation between these two sequences?

Since the time of Plato and Aristotle, it has been known that living organisms can be arranged in an ascending scale of organization (Scala Naturae). In the early nineteenth century, a number of embryologists observed that a rise of organization takes place during embryonic development, suggesting that in the course of ontogeny the embryo "climbs" the scale of nature. To those who at the time believed in Lamarck's theory, ontogeny would thus represent a recapitulation of phylogeny. This notion, relatively vaguely formulated in "Meckel-Serres law," is supposed to imply that we may expect to see a succession of adult forms make their appearance during ontogenetic development. This idea has survived to the present day, under the name of the "Biogenetic Law," given to it by its usurper, Haeckel (1866).

However, the law had in fact been refuted almost forty years earlier by von Baer (1828) in his renowned four laws. Confined for didactic reasons to the taxon Metazoa the generalizations embodied in the latter may be stated in modern terms in the following way:

Features distinguishing the taxa to which a metazoan animal belongs appear during ontogeny in a succession corresponding to the subordination of the several taxa.

This statement ("von Baer's theorem") has an implication which runs counter to a notion—shared, it seems, by almost all taxonomists—that evolution involves the transformation of organisms from one species to another. This idea, which represents a synthesis of Linnean systematics and the concept of evolution, was clearly stated by Darwin. Needless to say, this notion is not merely compatible with his micromutation theory and the biogenetic law; it is implied by them.

But this is not what we observe in the developing embryo. Rather, if we adopt the terminology of Linnean systematics and confine ourselves to the taxon Craniata, then the idea conceived by von Baer is that initially the craniate body plan is laid down, which is followed by, for example, the plan for a vertebrate, a gnathostome, a tetrapod, a bird, and a sequence of subordinate avian taxa. This succession corresponds approximately to the one we would follow if we entered the phylogenetic hierarchy at the level of the taxon Craniata and passed through it, to end up with some particular avian species. This coincidence may of course be regarded as completely insignificant; if it is not, then it implies that a kind of recapitulation of phylogeny occurs during ontogeny.

This recapitulation is not adult recapitulation as implied by the biogenetic law, nor is it really a true embryonic recapitulation, even if it is closer to the latter than to the former. I have previously suggested to apply the name "von Baerian recapitulation" to this phenomenon (Løvtrup 1978), but I shall argue here that it might as well have been called "epigenetic recapitulation," for the fundamental implication of von Baer's theorem is that in the distant past nature created the preconditions for the formation of the craniate body plan. This body plan was in the course of further ontogenetic development transformed to produce the first adult craniate animal. We do not know what it looked like, but both fossil and recent data suggest that it was a rather primitive organism.

In the course of time various modifications of this body plan have come into existence. Some of these have allowed for radical changes like the formation of jaws, lungs, and limbs, others have involved minor alterations. Whenever such a modification arose, the new organisms faced the same issue: survival or extinction. If they were exposed to competition with already existing forms, they would prevail if dominant—that is, more successful in the struggle for existence than the previous ones. If not, they would become extinct, unless for one reason or another they managed to become isolated and thus avoid competition. The second alternative implies a branch in the phylogenetic dendrogram. It thus appears that in contrast to the Linnean hierarchies, the Hennigian ones may be interpreted to correspond to events taking place in nature.

Epigenetics is the biological discipline which studies the various mechanisms responsible for the effectuation of ontogeny. Neglecting the philosophical difficulties involved in the concept of causation, we may say that epigenetics is the causal analysis of ontogeny, whereas classical embryology represents description of this process. Phylogeny and ontogeny merge, as it were, in the phylogenetic classification, which shows that there is a link between these two phenomena. And this link consists of the several epigenetic mechanisms which, for obvious reasons, must become engaged in the course of ontogeny in a succession corresponding to the one in which they have originated in the course of evolution.

Let us now turn to a discussion of some general aspects of epigenesis, following which I shall give a number of instructive illustrations of the mechanisms involved in epigenesis.

EPIGENESIS

Although epigenesis has been involved in phylogeny as well as in ontogeny, we can learn about this phenomenon only by studying development of living organisms. Some of the knowledge acquired through such endeavors is presented in this section, where I shall deal first with the subdivision of ontogeny into separate phases, and secondly with the epigenetic mechanisms.

Ontogenetic Phases

If the ideas developed above are correct, then it must follow that if it is possible to distinguish successive ontogenetic phases—an idea first suggested by von Baer and later elaborated by Sewertzoff (1931)—then each of them should correspond to a particular range of taxonomic categories in the systematic hierarchy, the rank of the taxa decreasing as development advances. This is a logical deduction, and it will be shown here to be corroborated by factual observations. Admittedly, the significance of this fact depends upon the number of phases which may be distinguished—the fewer they are the less the information conveyed.

Before this is done we must discuss a difficulty facing the outlined project. The concept of "categories," which is so handy in the present discussion, is a Linnaean convention that has little sense in phylogenetic classification because there are so many more levels or ranks in our example. However, if Linnaean taxa are monophyletic, they will be found in the phylogenetic hierarchy, and therefore the use of the conventional terminology in the present discussion will be correct to a certain extent. Yet, it must be realized that taxa, which are now ranked at the same categorical level, may have widely different ranks in the phylogenetic classification. Furthermore, the number of levels will vary greatly from one region of the hierarchy to another.

With this reservation I may proceed with my analysis of ontogenesis and observe first that in many cases this phenomenon is interrupted by one or more instances of metamorphosis. These may amount sometimes to moderate and sometimes to profound morphological modifications, and generally they are also followed by changes in the ecological habitat. Metamorphosis involves a considerable complication of ontogenesis compared with those simpler cases dealt with here. However, if it is assumed that metamorphosis represents a second phase of form creation, as defined below, then I believe that the generalizations presented here may still be valid.

Ontogeny can be subdivided into a progressive and a regressive period, the former most often lasting from fertilization to maturation, and the latter stretching from thereon until death. The progressive period may be subdivided into various phases, the number of which depends entirely on the chosen criteria. In general, it seems convenient to distinguish four different phases: cleavage, form creation, differential growth, and allometric growth.

Cleavage

Before true morphogenesis sets in, the fertilized egg undergoes a number of cell divisions, often distinguished by being synchronous and very rapid.

In most animals this event transforms the solid egg into a hollow sphere, the blastula. Exceptions to this generalization are common, as exemplified, for instance, by the development of insects, and of amniotes with large yolk-laden eggs. This topological transformation is in itself an event of morphogenesis, and I have previously included it in the phase of form creation. However, when ontogeny and phylogeny are to be compared, then it is expedient to consider cleavage as a separate ontogenetic phase.

The reason for this is that if we are to operate with von Baerian recapitulation, then we must have a starting point. Now, as I shall discuss below, it may be claimed with some justification that the formation of the body plan of the highest taxa within the subkingdom Metazoa, generally phyla, begins with the onset of gastrulation. Therefore, if evolution from one phylum to another were to be recapitulated during ontogeny, then this should occur during the phase of cleavage. No traces of such recapitulation may be observed, as is perhaps understandable, considering that the principal characteristic of the various metazoan phyletic body plans is that they have nothing but very general features in common, for instance bilateral symmetry or larval forms. Under these circumstances the wide differences observed between the process of cleavage in the various phyla is of subordinate significance, so that this phase may considered simply as one of necessary preparation for morphogenesis to take place.

Form creation

Thus, true form creation begins with the process of gastrulation. Formally, this event involves the transformation of a single-walled spherical blastula into a double-walled spherical gastrula. In the most primitive animals this event takes place through ingression of cells, but in most cases it involves the topological transformation of the blastula through invagination. From the present point of view it would be highly desirable if the process of gastrulation was homologous in all metazoan taxa, but it appears that the processes which initiate the formation of the germ layers in certain teleosts and amniotes not even formally can be equated with the normal process of gastrulation (Ballard 1981). According to the traditional recapitulation theory this is most unfortunate, but if, as suggested here, recapitulation begins only during gastrulation, such differences are less important.

What happens during the phase of form creation is that the phyletic body plan is laid down, consisting in the taxon Craniata of a notochord, enclosed above by a spinal cord ending in a brain at the anterior end, and at both sides by a row of metameric somites, and below by the gut. This body plan subsists unchanged in the taxon Myxinoidea—the hagfishes. If cartilage-producing cells engage in the formation of neural arches around the spinal cord, the vertebrate body plan obtains, the first traces of which are found in Petromyzontidae, the lampreys. This circumstance has the consequence that Petromyzontidae ought to be united with Gnathostomata in the taxon Vertebrata (Løvtrup 1977; Janvier 1981). If cartilage-producing cells in the head, formally representing the skeleton of the first gill arch, accumulate to form the primordia of the jawbones, the gnathostome body plan is realized. Without going into detail as far as the subsequent steps are concerned, it may be mentioned that if limb buds are formed, the embryo will develop into a tetrapod; otherwise it will become a fish. If we disregard the diagnostic significance of the embryonic membranes, shortly afterwards it can also be determined visually to which of the major tetrapod taxa the organism belongs.

Some ambiguity obtains with respect to the exact extension of the phase of form creation, particularly when we go beyond the taxon Craniata. However, in all of those cases where no metamorphosis occurs, the developing animal at the end of this phase embodies all the distinguishing features of the major taxa (phylum and class) to which it belongs. However, morphologically it exhibits a general type, which does not allow for any predictions about its further development (fig. 7.1). It may be observed that the use of the names "phylum" and "class" here involve unavoidable inaccuracies.

Differential growth

If all morphogenetically significant parts are present—this holds mostly, but not exclusively, for the skeleton—then the further development ensues by a process of differential growth. This process involves the determination of the absolute size of the animal, but above all it establishes those

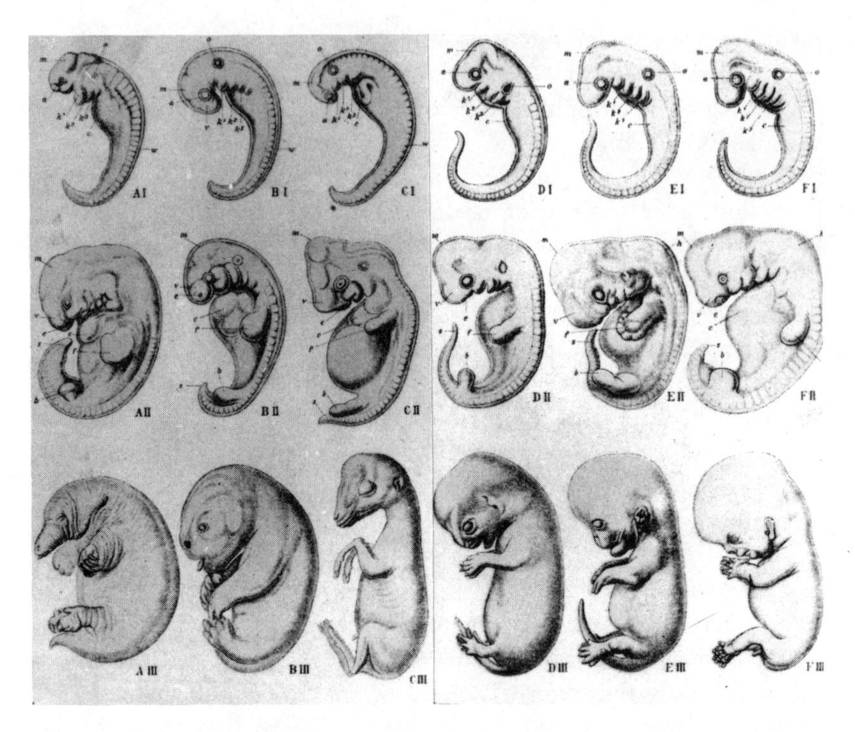

Figure 7.1. Vertebrate embryos at different stages of development. The transition between the phases of form creation and differential growth occurs sometime between the first and second row. *A*, *Echidna* (spiny anteater); *B*, *Phascolarctos* (marsupial); *C*, *Cervus* (deer); *D*, *Felis* (cat); *E*, *Macacus* (monkey); *F*, *Homo* (man). From Haeckel (1920).

particular morphological features which characterize the lower taxa to which the organism belongs, down to the generic or specific level.

Allometric growth

Once again it is in many cases difficult to decide when the transition occurs from one phase to another. However, very often the transfer from the phase of differential growth to allometric growth may be considered to coincide with hatching or birth, or else with metamorphosis. Usually the specific affiliation is evident at this time, but in some instances—e.g. small birds and mammals—the morphological similarities between different but related species are so great that the specific affinity can be established only at a later stage of development. However this may be, there are typical differences in relative form between young and mature animals, and these are brought about during a growth process, because it is allometric. There is hardly any principal difference between differential and allometric growth; the main reason for making the present distinction is that the earlier growth processes, although slighter in extent, are much more divergent.

Epigenetic mechanisms

Three different kinds of epigenetic mechanisms may be distinguished: cell division, cell differentiation, and morphogenetic processes.

Cell division

The process of cell division comprises two separate events, karyokinesis and cytokinesis. The former is accomplished by microtubules, assisted by the centrioles, the latter by actin filaments located in the cortex. To account for the activities of the various agents involved in this physical process is of great interest, but beyond our scope here. I shall mention only two points where the process of cell division is directly involved in epigenetic events.

First, as noted above, in many taxa the cell divisions occurring during the phase of cleavage are synchronous and very fast. The number of divisions (10 to 15) is strictly controlled under normal conditions. Various observations suggest, however, that it is not the number of cell divisions as such but the amount of DNA incorporated in the nuclei which determines the extent of the phase of cleavage. Since there is a store of deoxynucleotides in the fertilized embryo which corresponds fairly well to the number of cells formed during this phase, it has been proposed that this store may be a decisive factor (Løvtrup 1974). Yet, other agents which, like deoxynucleotides, are used in regular amounts at each cell division may also be involved (Newport and Kirschner 1982).

Second, in the classical embryological literature one may encounter the claim that cell differentiation can occur in the absence of cell division, as exemplified by ciliation occuring in an unfertilized egg of *Chaetopterus* (cf. Needham 1942). It may be questioned whether or not this truly represents cell differentiation, but I think that it would be wise to define this phenomenon unambiguously so that the mentioned situation does not arise. As a proper definition I suggest that cell differentiation occurs only in Metazoa, and only when at least two different patterns of differentiation are present in the same organism. By a pattern of differentiation I do not mean merely that the cells are visually dissimilar; e.g. in the amphibian embryo a large vegetal yolk-laden cell may represent the same pattern as a small animal cell, filled with melanosomes. The essential point is that the cells representing a particular pattern all transcribe the same part of the genome, and that this is different for each pattern.

From this definition it clearly follows that cell division is a necessary, but an insufficient precondition for cell differentiation. Since, according to the generally accepted tenet, the genome is the same in all cells, it has been inferred that the differential transcription implied by the definition is controlled by extranuclear, or perhaps even extracellular, factors. If this is true, then it evidently follows that cell divisions unequal with respect to the decisive factors are a necessary prerequisite for cell differentiation. If desired, "unequal cell division" may be defined such that it even becomes a sufficient precondition.

Cell differentiation

Cell differentiation is the basis of the distribution of work which characterizes the metazoan body. By the singular transcription of a certain part of the genome the cells belonging to a particular pattern of differentiation are enabled to carry out a special function. Note that the part of the genome required for this specialized activity probably is very slight compared to the basic cell repertoire required by all cells (Britten and Davidson 1971).

A basic precondition for cell differentiation is the occurrence of unequal cell divisions, involving perhaps that the daughter cells from a particular division differ with respect to size or cytoplasmic composition, or with respect to the environment. Under such circumstances it may happen, if the cells are in a state of competence, that the two sister cells differentiate along separate paths. The inhomogeneities responsible for the unequal cell divisions may be referred to as "polarities" (Løvtrup 1983).

The possibilities envisaged here play a large role during the earliest stages of development, but later on other mechanisms take over, among which may be mentioned induction, autodifferentiation, and inhibition.

Induction is an instructive event—that is, it somehow changes the programming of a cell. In some cases cells may not spontaneously begin the differential activity for which they are programmed. A special kind of stimulation may then be required, for which the word "activation" seems to be appropriate. It is interesting to note that according to current views the process that leads to the formation of the central nervous system in the amphibian embryo is an induction; the ectoderm is presumably programmed to form epidermis. It seems likely, however, that the opposite relation holds—that is, the ectoderm is programmed to form nerve cells but must be activated in order to realize this pattern of differentiation. If this does not happen, it may be induced to form epidermis, probably under the influence of the external medium (Løvtrup et al. 1978; Løvtrup, 1984).

Two kinds of induction, homotypic and heterotypic, may be envisaged. The former implies that a cell imposes its own particular pattern of differentiation on a "naive" cell. This mechanism seems to be of great importance during embryonic development, and may indeed be the one responsible for tissue homogeneity in the metazoan organism. Heterotypic induction must imply that a cell of a special kind forces another cell to differentiate in a direction different from its own. Clearly, this cannot be a general mechanism, for in that case tissue homogeneity would be difficult to achieve.

So far as activation and induction are concerned, it is reasonable to presume that they are exerted by chemical substances secreted by the cells or residing on their surfaces. It may be presumed that activating substances have little specificity, while inductors must specify either the entire pattern of differentiation or one regulative event which in turn sets off the process. It has been found *in vitro* that enzyme substrates, when present in the medium, may function as inductors in amphibian embryonic cells, but whether they do so *in vivo* is not known (Stern and Kostellow 1958; Løvtrup et al. 1984).

Turning to the mechanism of autodifferentiation it may be observed that amphibian ectodermal cells isolated *in vitro* and properly "induced" will differentiate into a series of patterns among which mesenchyme cells, nerve cells, and melanophores have been identified, and it is possible that even glia cells are formed. The various patterns of differentiation seem to be independent of each other, being determined only by the time at which they arise. This series of patterns clearly may be characterized as a neural crest, or rather a trunk neural crest repertoire. There is another series which has often been observed, involving perhaps mesenchyme cells → collagen producing cells → chondrocytes → osteocytes. This series is a typical mesodermal series, but may also be characterized as a cranial neural crest repertoire (Hörstadius 1950; Weston 1970).

One might hypothesize that metabolic products from one stage further the passage to the next one, but if so, it would seem difficult to avoid the conclusion that all cells end up at the final stages. A more likely interpretation is that a mechanism of exhaustion is involved in which metabolites, which are consumed in a particular succession, permit certain differentiations to occur, while inhibiting the subsequent ones. Such a mechanism would be possible only in the early embryo, before circulation has become established. It is typical that such sequences of cell differentiation are observed only in embryonic cells, and perhaps only when they are cultured in a medium that is far from being "complete."

To the extent that cells may act as inductors, whether homotypic or heterotypic, they evidently also function as inhibitors—namely, with respect to other differentiation patterns. It is likely that inhibition in many other instances is of importance for cell differentiation, but this topic cannot be dealt with here. One kind of inhibition should be mentioned: inhibition of existence, or cell death. As we shall see presently, cell death is a very important morphogenetic agent.

Cell differentiation in embryos is typically distinguished by the gradual appearance of new patterns of differentiation. It is probable, and in line with the ideas outlined above, that the basic arrangement of these patterns constitutes a bifurcated dendrogram, but with allowance made for interactions between the several differentiation patterns this simple model soon becomes very complex.

It seems possible that the first few cell differentiations may constitute basic cell types, or cell orders, to which each of the many patterns may be accorded. The first attempt to distinguish such fundamental cell types was made by Willmer (1960). On this basis I have proposed that there are eight fundamental cell orders, whose relations may be represented in a "cell cube" (fig. 7.2). Thus it is suggested that morphologically the egg cell and the early blastomeres are amoebocytes, distinguished by the absence, during the interphase, of microtubules and perhaps also of in-

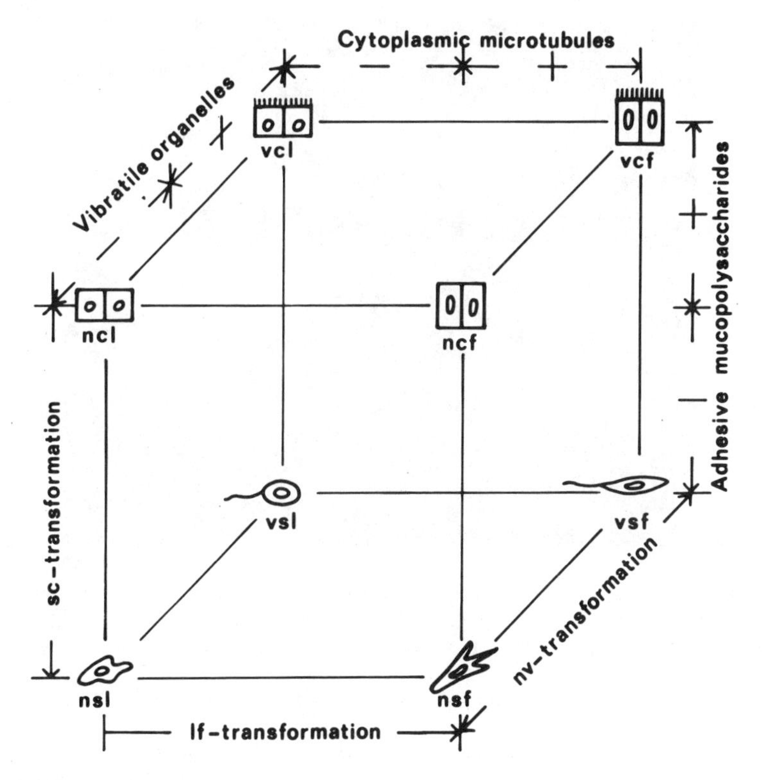

Figure 7.2. The cell cube. Representatives of eight cell orders are shown in the corners of the cube. Each median plane divides the orders in two classes, distinguished by the presence or absence of one or more features. To "adhesive mucopolysaccharides" should be added "etc." From Løvtrup (1974).

termediate filaments. Having no cytoskeleton the cells are fundamentally spherical, but they possess an ectoplasmic cortex containing microfilaments. The latter may be used to form filopodia or lobopodia, which enable the cells to migrate when exposed to a proper substrate. For convenience these cells have been called "solo-lobocytes" (sl-cells) (Løvtrup 1974, 1983).

The sl-cells may in turn differentiate in two separate directions, a dichotomy that is beautifully modeled by the fate of the amphibian ectodermal cells, that is, either induced epithelial cells, as represented by the epidermis, or activated nerve cells and other patterns of differentia-

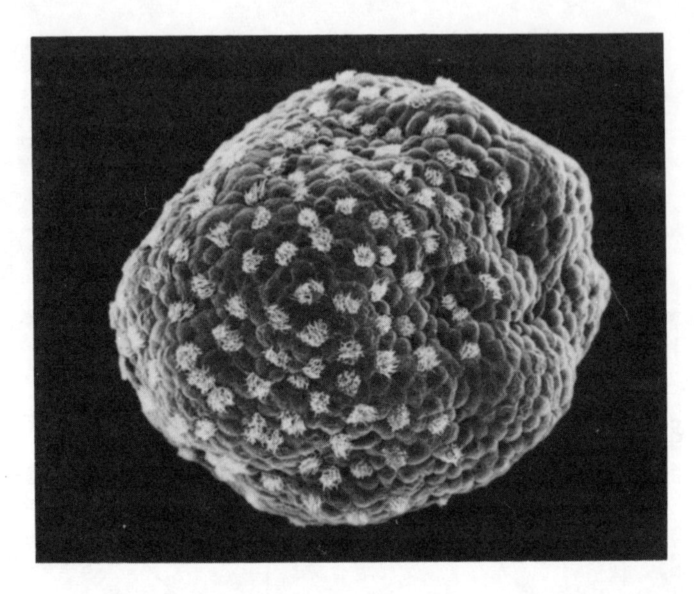

Figure 7.3. Spherical vesicular aggregate of ciliated and nonciliated epidermal cells formed by ectodermal cells isolated from the axolotl blastula (75 ×). From Løvtrup-Rein, Landström and Løvtrup (1978).

tion. In my interpretation epithelial cells, or "colligocytes," are distinguished by the formation of epithelia, an ability which may depend primarily upon the adhesive properties of the cell surface. However, on top of this, epithelia are stabilized partly by junctional specializations of various kinds, partly by a basal lamina in which is incorporated a layer of reticular collagen (type IV), the macula densa. Since this colligocyte is derived directly from the solo-lobocyte, I have suggested the name "colligo-lobocyte" (cl-cells) for this order of cells (fig. 7.3).

The other derivative is a free cell—a solocyte—which, in contrast to the sl-cell, contains a rigid cytoskeleton, made up of intermediate filaments and microtubules. When unattached, this cell may still be spherical, but when exposed to an appropriate substrate, it will extend filopodia which become attached to the available surfaces, and the cytoskeleton then contributes to the flattening and the shaping of the odd contours of the cell. The latter, which to the general mind is represented by the fibroblast (fig. 7.4), has been called a "solo-filocyte" (sf-cell), notably because it can extend long filopodia containing microtubules.

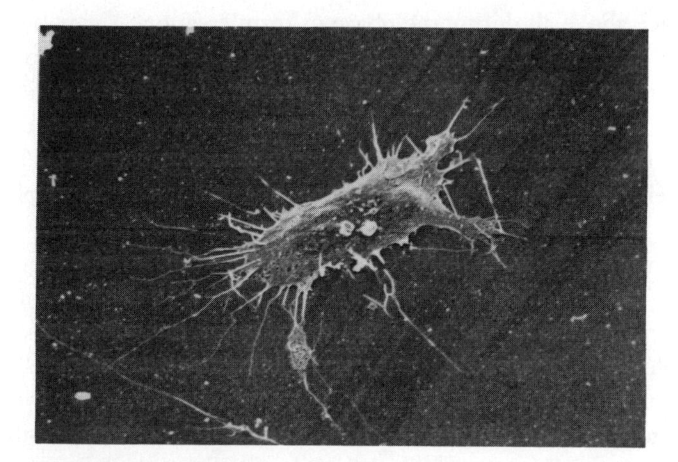

Figure 7.4. Solo-filocyte formed by axolotl embryonic ectoderm after "induction" with LiCl (750×). From Løvtrup et al. (1978).

Finally, a cell type may exist which incorporates both the properties of the cl-cell and the sf-cell, the "colligo-filocyte" (cf-cell), which, as may be predicted, is an elongated cell present in epithelia of various kinds, represented typically by the cells in the neural plate.

Corresponding to each of these four cell orders is a cell possessing vibratile organelles, either cilia or a flagellum, the former being confined to the colligocytes and the latter to the solocytes. A discussion of these cell types lies beyond the scope of the present paper.

As we shall see in the next subsection, these cell types, notably the cl-cell and the sf-cell, are of utmost importance for morphogenesis. The latter is particularly distinguished by its participation in various dynamic activities. Furthermore, to a large extent products secreted by various kinds of differentiated sf-cells form the skeleton, and hence contribute significantly to the body form.

Morphogenetic mechanisms

We have already discussed the morphogenetic event which follows from the occurrence of the cleavage divisions—that is, the transformation of a solid sphere into a hollow sphere. Before we pass on to discuss the mechanisms responsible for the various morphogenetic processes, which take place above all during the phase of form creation, it may be to the point to bring out the implication of morphogenesis. This phenomenon is a spatiotemporal transformation—to a certain extent topological—of a hollow sphere. It is a physical process and it may therefore be premised that if we are going to understand morphogenesis, then we must look in the developing embryo for agents which are able to carry out physical work.

This inference seems to imply a refutation of two ideas in current biology, namely, that ontogenesis is programmed in the genome, and that there are such substances as morphogens which can be made responsible for more or less extensive aspects of morphogenesis. As is well known, the genome contains a digital program, specifying the synthesis of a large number of RNA molecules, which in turn are responsible for the synthesis of the proteins which may be produced by the organism in question. There cannot be any doubt that many of these proteins are involved in morphogenesis, even to the extent that they develop the physical forces which carry out the morphogenetic processes. But that does not imply that the information contained in the genome specifies morphogenesis. Anyone who questions this proposition may test it by dissociating an embryo at any stage of development; if this is done, and the cells are supplied with the proper nutrients, they may survive and perhaps differentiate according to a certain program, but all morphogenesis comes to an end.

Clearly, the organization of the embryo on one or more supracellular levels of organization is of importance for morphogenesis. It may be mentioned that the basis of this organization lies in the information embodied in the egg cell and the centriole, which together account for the process of cell division. In contrast to that resident in the genome, this information is analogic.

As far as the expression "morphogen" is concerned, I am convinced that there are many chemical substances which deserve to be classified under this heading—such proteins as collagen, glucosaminoglycans, glycoproteins, etc. But the expression is used much more in a metaphysical sense, generally referring to substances supposed to be present in the cytoplasm of the egg or the early embryo, and distributed in such a way as to account for one or more of the polarities. The existence of asymmetric distributions of this kind seems very unlikely, since they might become upset by diffusion, or by the various dynamic events (contractions, etc.), which occur in association with fertilization. Yet, even if this point were granted, it is still highly dubious that the substances in question could be morphogens; it is much more likely that they affect the differentiation of the cells. The circumstance that such events in turn may influence morphogenetic processes does not warrant their given name.

After this digression let me try to analyze the morphogenetic mechanisms. First I might consider aggregate formation, by which I understand permanent formations, e.g. epithelia. This task is performed by the colligocytes. The aggregates may take the shape of tubes, vesicles, etc. Topologically, it might well be possible to analyze the metazoan body in terms of tubes and vesicles with one exception, the notochord, which is a solid "epithelial" aggregate.

The other kind of morphogenetic activity is the execution of work, the prerogative of the solocytes. In the amphibian embryo the process of gastrulation seems to be carried out by sl-cells, but subsequently it is probably the function of the much more motile sf-cells which are involved in the various dynamic activities. Clearly, it is the motility which is the precondition for this activity, because the physical power which is responsible for the movement of single cells may under other conditions become involved in the topological transformation of epithelial aggregates, a common event during early morphogenesis. In many such instances it is possible to demonstrate the participation of sf-cells.

The sf-cells occur in two situations which superficially may resemble aggregates as formed by colligocytes. The first is represented by capillaries, which seem to be formed by flattened sf-cells (endothelial cells), held together through contact inhibition. That capillaries are not true aggregates is demonstrated by the fact they are labile, undergoing constant breakdown and reconstruction. The other situation is the aggregations which form the rudiments of endoskeleton. These are formed by sf-cells, mesenchyme cells, which differentiate to produce an extracellular matrix, thus serving to keep the cells together.

This last observation may call attention to the other morphogenetic element, extracellular substances, which serve to stabilize cell aggregates and to cause swelling through uptake of water, thereby creating the conditions required for various cells to exert their morphogenetic activities.

We have seen here that cells are very important morphogenetic agents. But cells are small, and they can work only over very short distances.

This is why all the really important morphogenetic events must occur when the embryo is very small, when it still has cellular dimensions.

The Workings of Epigenesis

In order to illustrate how the various epigenetic mechanisms exert their function we shall here discuss a selected variety of practical examples, subdivided with respect to the ontogenetic phases.

Cleavage

On the cellular, in contrast to the molecular, level the process of cleavage does not seem to offer any great problems. There are some points, however, which deserve mention. The cells formed through the cleavage divisions must be kept together in order to form a blastula. I believe that many embryologists consider the blastula wall to be formed by epithelial cells, that is, by cl-cells. It is easy to demonstrate that this cannot be true. Thus, the basal lamina typical of colligocyte aggregates is missing; in fact, if it were not, the cells would be unable to carry out the process of gastrulation. Rather, as shown first by Holtfreter (1943), the cells are kept together by a membrane—or surface coat—situated at the outside of the blastula. This membrane consists of a protein or glycoprotein, which may be removed,—resulting in the scattering of the cells—by various expedients that do not affect epithelial aggregates (Mattsson et al., in press).

In the present context I might also mention the question about the total number of cells in the adult body, a point which possibly may be related to the cell divisions going on during cleavage. Cell size varies considerably from one major taxon (phylum or class) to another, but within these it usually is quite constant. This means that body size in general must be proportional to cell number, and this in turn seems to imply that body size is controlled by the number of cell divisions. If this is true, then body size within closely related organisms ought to vary by a factor of 2.

It is known that single mutations may cause a doubling or a halving in body size, and several observations show that this effect is related to the number of cell divisions during ontogenesis (rabbits, Painter 1928; Castle and Gregory 1929; insects, Raff and Kaufman 1983). In fact, there are many data in the literature suggesting that this is the way that body size changes commonly occur, at least in the higher vertebrates (Løvtrup et al. 1974; Løvtrup and Hansson Mild 1979). However, the available information is insufficient to obtain statistically significant answers on this question (Roff 1977).

This does not imply that cell number is the only factor in control of body size. Hormones, which are intimately involved in the normal growth processes, may in many instances affect the final body size, often with disharmonic consequences.

Form creation

As we have seen, the task facing the epigeneticist is to give a causal explanation of the various events taking place during ontogeny. It may perhaps be claimed that this is exactly what the majority of current developmental biologists occupy themselves with when they study various macromulecular and cellular aspects of this phase of development. I shall contend that although these efforts may be a valuable contribution to our understanding of the mechanisms involved, it nevertheless holds that they cannot teach us much about the true processes of form creation, for this phenomenon takes place only in whole embryos, and must be studied on whole embryos. From this standpoint I shall discuss here some of the events which occur during the phase of form creation, citing examples taken primarily from the amphibian embryo, the one with which I am most familiar.

The intimate connection between cell differentiation and morphogenesis becomes apparent when we study the mechanisms involved in gastrulation. Of the two basic cell differentiations discussed above, the formation of sf-cells begins "spontaneously" in the endoderm in a region approximately midway between the equator and the vegetal pole. We do not know what determines this particular location, but we may formally ascribe it to an animal-vegetal polarity.

An interesting and very important fact is that the appearance of the sf-cells does not take place simultaneously around the circumference of the blastula, it begins at the dorsal side and spreads at each side in two waves which meet at the ventral side of the egg. This temporal gradient imposed on the process of differentiation is a topological precondition for gastrulation.

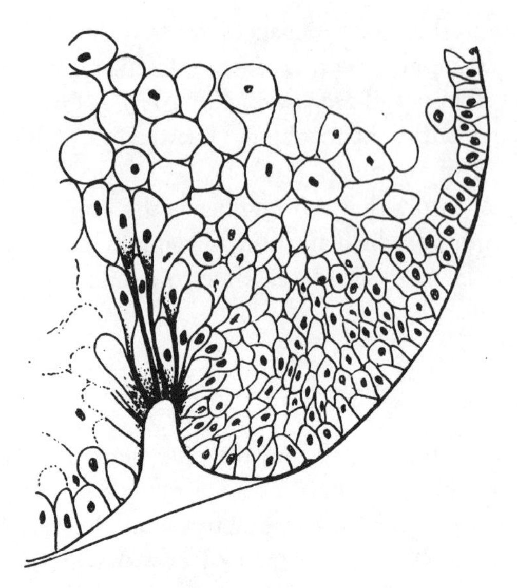

Figure 7.5. Sagittal (quasi-median) section through the blastopore region of a urodele gastrula showing how the vegetal sf-cells are pulling in the surface of the embryo. From Balinsky (1981).

In contrast to the original sl-cells, the sf-cells are able to extend long filopodia which pull the surface of the blastula into the interior, thereby forming the blastopore, outlined by dark pigment (fig. 7.5). As a result of this activity, two layers of membrane-bound ectodermal cells are brought into contact. Since these cells are motile, those of the lower layer begin to migrate in the direction where space is available—that is, toward the animal pole. However, the ground on which they tread, the outer cell layer, is not *terra firma;* rather, its toughness is comparable to that of the layer of migrating cells, and therefore it yields, so that the outer cells advance downward while the inner cells go upward. If this did not happen, the product of gastrulation would be a double-walled hemisphere rather than a sphere. As the process of invagination continues, the blastopore will spread around the embryo, forming first a crescent, later a large circle and finally a small circle.

From the fate map we know that the cells which end up in the interior are the presumptive notochordal and the endodermal cells. Among these we must look for those responsible for carrying out the work, those

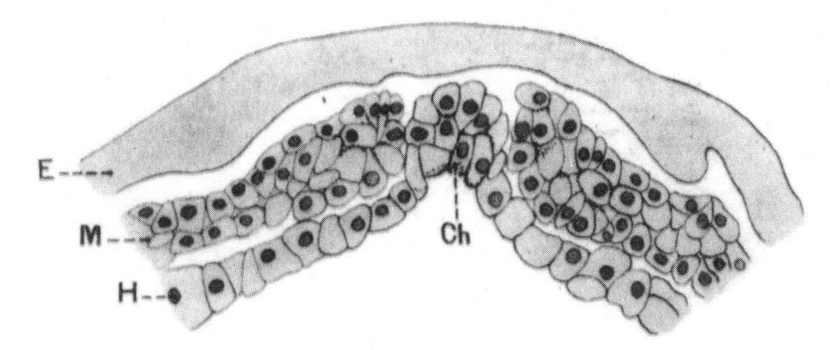

Figure 7.6. Transverse section through an axolotl gastrula, showing (1) that only the presumptive notochordal cells in the dorsal median establish contact with the exterior ectodermal cells, (2) that the notochordal primordium is associated with the endoderm, and (3) that the latter is separated from the ectoderm by a layer of mesodermal cells. *Ch*, notochord; *E*, ectoderm; *H*, endoderm; *M*, mesoderm. From Brachet (1921).

involved in the process of gastrulation. Superficially, one would hardly expect the large endodermal to perform this task, and this suspicion is easily confirmed when we make a cross section of the gastrula (fig. 7.6). It is seen that contact between two layers of membrane-bound cells obtains only around the dorsal median, where the inner layer represents the presumptive notochord. To both sides of the latter the endodermal cells are separated from the ectoderm by a layer of free cells, the mesoderm, which prevents them from establishing contact with the ectoderm. Because the notochordal cells alone carry out the work, they undergo a considerable topological deformation, and come to lie as a narrow band around the dorsal median (fig. 7.7). At the end of gastrulation the association between notochord and endoderm comes to an end; the resulting cleft in the endoderm is closed by cells (presumably sf-cells), which pull the edges together.

We must now return to the question of cell differentiation. It turns out that all the ectodermal cells—except the neural crest cells—become colligocytes. The reason why they do so is not known, but there are various observations which suggest that exposure to a dilute medium, perhaps also to low concentrations of Calcium ions is a necessary prerequisite. Innumerable experiments have shown that under these conditions the ectoderm differentiates into ciliated epidermis. However, in

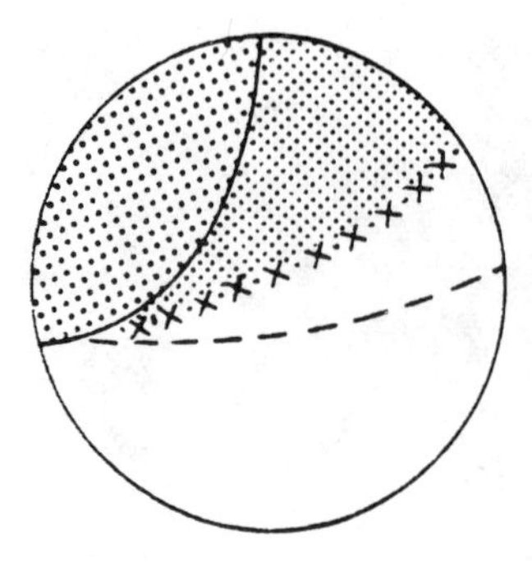

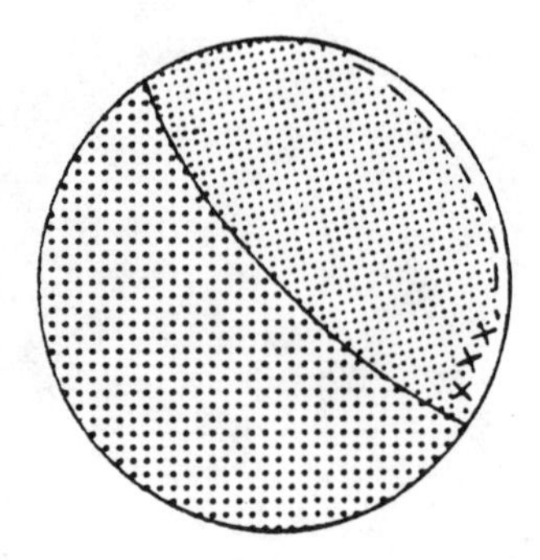

Figure 7.7. Diagram of the topological transformations occurring during gastrulation in the urodele embryo. Coarse stipple, epidermis; fine stipple, neural plate; x.x.x, limit of invagination; —, limit between notochordal primordium and endoderm. The deformation of the notochordal primordium is indicated, but it should be recalled that it is situated below the neural plate, and thus not visible from the outside. From Løvtrup (1983).

the embryo only the ventral ectoderm becomes epidermis, while the dorsal cells differentiate to form neural plate and notochord. We know that this happens under the influence of the invaginating sf-cells, and it is reasonable to presume that the effect of the latter is to ensure that the ectodermal cells become filocytes, and hence colligo-filocytes. This inference is confirmed; the cells in the neural plate assume the elongate shape typical of this cell order, as we shall see presently.

The situation is a little more complicated with respect to the notochordal cells, primarily perhaps because they form a "one-dimensional" aggregate. However, when differentiation of these cells is followed, it is quite obvious that they become colligocytes, for they establish very close intercellular contact with each other, and furthermore surround themselves with a layer of reticular collagen so typical of this cell class. Furthermore, they undergo considerable swelling—a circumstance that may be correlated with the fact that they begin to produce hyaluronate, which is retained by the cells (fig. 7.8). The oncotic pressure thereby produced is counteracted by the tension in the surrounding collagen network, and together these forces give a certain mechanical strength to the notochord, enabling it to function as an embryonic skeletal element. The cell differentiation may itself cause a certain elongation of the notochord, but during the subsequent development, the process continues because new

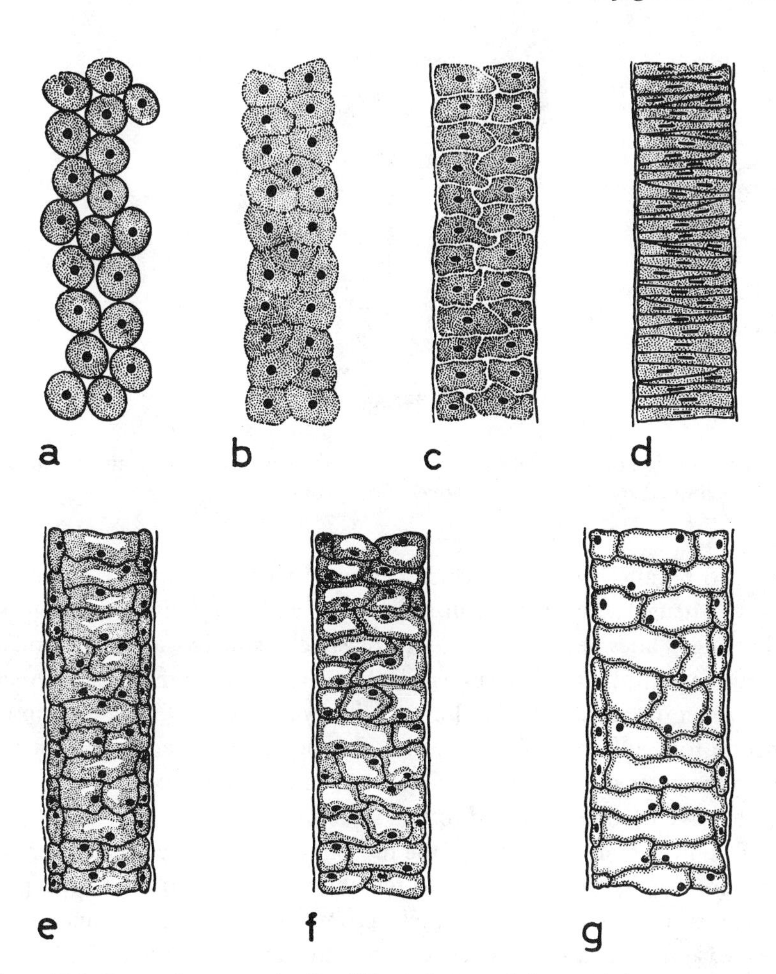

Figure 7.8. Successive stages in the differentiation of the amphibian notochord. (a), late gastrula; (b), early neurula; (c) neural plate; (d), early tailbud; (e), middle tailbud; (f), late tailbud; (g), larva. From Mookerjee, Deuchar and Waddington (1953).

cells are incorporated at the posterior end as long as any are available for this purpose.

At the end of gastrulation the exterior of the embryo is taken up by the presumptive epidermis at the lower side, and the presumptive central nervous system at the upper side. Initially these two primordia are indistinguishable, but very soon the neural plate becomes distinctly outlined

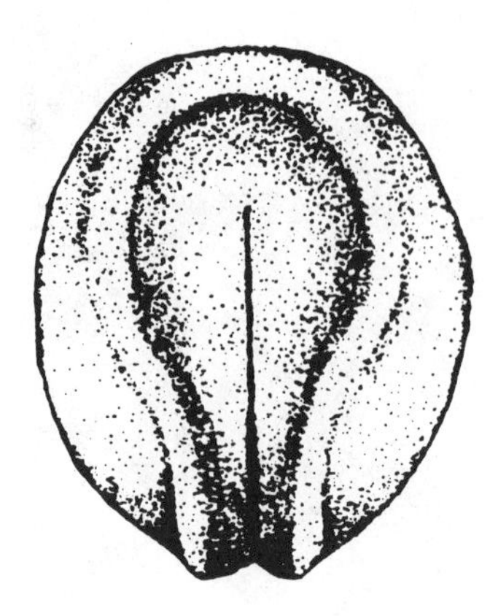

Figure 7.9. Urodele neurula showing the median neural groove and the neural fold delimited by two neural grooves. From Rugh (1948).

by two parallel dark lines which delimit the neural fold; the embryo is now a neurula (fig. 7.9). Our experience with the blastopore suggests that these lines reveal the activity of sf-cells situated at the boundary between neural plate and epidermis. We shall later see that this inference is correct, because the neural crest cells, typical sf-cells, originate in the neural fold. Of particular interest is the median groove, for it coincides with the extension of the notochord, and indicates that cells in the neural plate have become attached to this structure.

Two events are decisive for the morphogenesis of the central nervous system. The first is that, as indicated above, the cells in the neural plate become colligo-filocytes. As they become elongated, their apical ends decrease in size, and as a consequence the area of the neural plate decreases. At the same time the notochord begins to stretch, slightly anteriorly, and extensively posteriorly (fig. 7.10). This in itself causes a considerable narrowing of the part underlain by the notochord; it also seems to have one further consequence—the edges of the neural plates are raised (Jacobson 1978). This proposition is supported by a mechanical model. When a piece of rubber plate is stretched, two folds will be raised, the heights of which increase with the stretching. In the neural plate this process is often most extensive at the transition between brain and spinal cord, that is, where the notochord stretches in both directions. If Jacobson's theory is correct, it may be no coincidence that it usually is here that the two edges of the neural plate first meet and fuse. Once contact

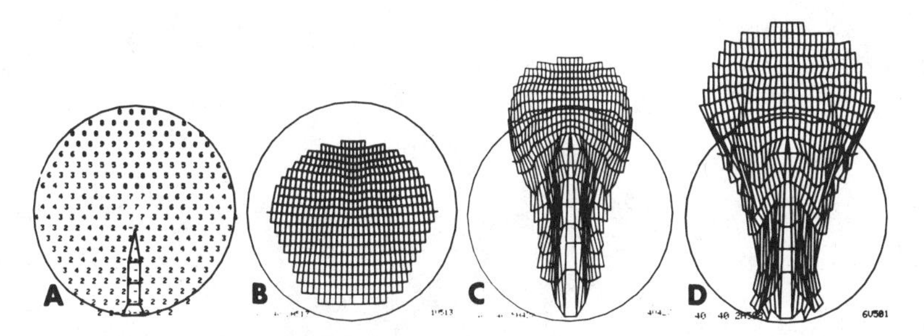

Figure 7.10. Computer simulation of factors influencing the topological deformation of the amphibian neural plate. (A), initial stage; (B), cell differentiation—cell elongation and shrinkage of surface area; (C), cell differentiation and notochordal stretching; (D), notochordal stretching. From Jacobson (1978).

is established, it is easy to understand how the process of fusion might spread to the other regions of neural plate. It is interesting to observe that the attachment of the neural plate to the notochord is a precondition for the existence of the craniate brain.

Returning now to the mesodermal cells, it has been found that, except for the dorsal and the most anterior region, these are intercalated between the ectoderm and the endoderm. This implies that the mesodermal cells advance together with the endoderm. It is possible that the mesodermal cells are passively dragged along, but they are able to migrate autonomously, as indicated particularly by the behavior of the cells at the anterior edge of the mesodermal mantle (fig. 7.11). When the neural plate narrows, the ectoderm is pulled toward the dorsum, and some mesodermal cells are dragged along and become accumulated as two elongated aggregates at both sides of the notochord (cf. fig. 7.6).

The differentiation of the mesoderm proceeds in three basic directions, a phenomenon which it seems possible to account for, in crude terms at least, on the basis of the suggested cell classification. Thus, the cells situated next to the notochord and the neural tube become colligo-filocytes under the influences of the mentioned structures. This does not explain that they subsequently become muscle cells; the difference between the ectodermal cf-cells which become nerve cells, and the mesodermal ones, must be a consequence of the separate origin of these two cell populations.

As the dorsal mesoderm differentiates into colligocytes (cf-cells), it has

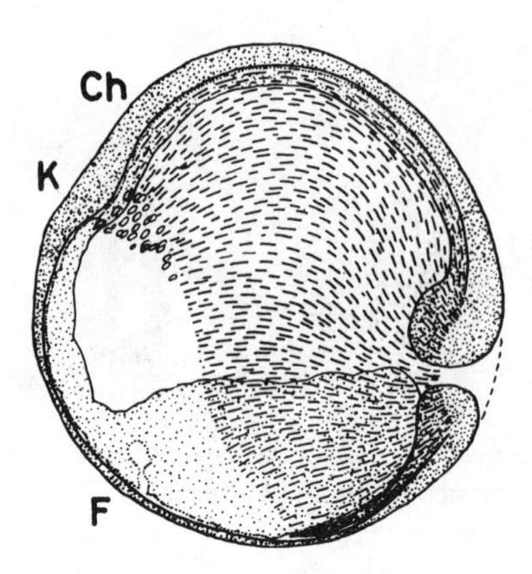

Figure 7.11. Migration of the mesoderm. Median section with reconstruction of the mesodermal layer in the early urodele neurula. Ch, presumptive end of the notochord; K, anterior end of neural plate. From Vogt (1929).

the same consequence as may be observed when colligocytes differentiate *in vitro*—namely, the formation of vesicles (fig. 7.12). As this process occurs in an antero-posterior direction, a series of vesicles, the somites, is formed at each side of the notochord (fig. 7.13), each surrounded by the distinctive collagenous network of the basal lamina. This process of differentiation is the basis of the metameric segmentation of the craniate body; all further metamery in the trunk is causally dependent on that of the somites. Thus, the vertebrae are formed around the spinal cord, one between each pair of somites, and this metamery in turn determines that of the central nervous system. The sympathetic and spinal ganglia are formed mainly by neural crest cells which, migrating in the collagen network between the epidermis, the spinal cord, and the notochord, occupy spaces between the somites (fig. 7.14). Finally, the ribs are laid down in the spaces between two somites and the skin.

The mesodermal cells situated below the somites in the interior differentiate into sclerocytes, contributing to the formation of connective tissue. These cells are initially free cells, being typical sf-cells. On the other hand, the cells situated below the ventral epidermis form epithelia which enclose the body cavity, the coelom; their fate suggests that they are colligo-lobocytes.

These are the most important events involved in the morphogenesis of

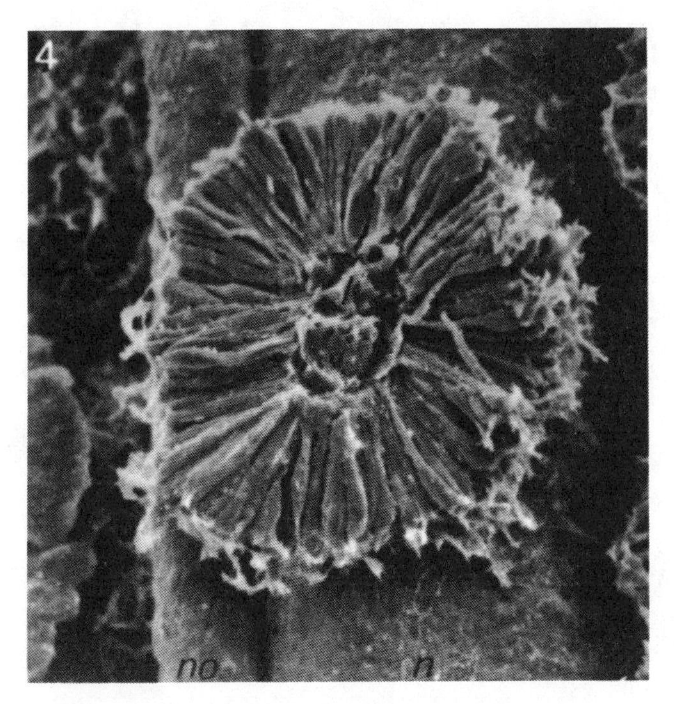

Figure 7.12. Sagittal section through a somite at the stage 11 chicken embryo. The somite is seen to be a vesicle made up by elongated cells. *no*, notochord; *n*, neural tube. From Bellairs (1979).

the trunk, but at the same time the heart and the blood circulation is established, and the endoderm forms the digestive tract, processes of great significance, particularly for body function. We shall not discuss these phenomena here, nor shall we deal with the intricate morphogenetic processes taking place in the head.

However, one point may be worth mentioning—the cause of the difference between the morphogenesis in the head and in the trunk. According to the rules laid down it must be possible to trace this difference back to cell differentiation processes. Now, the morphogenesis of the head region is characterized by the extreme swelling which takes place there. This we may ascribe to the production of excessive amounts of hyaluronate, presumably by the so-called "mesenchyme cells." In the trunk the origin of mesenchyme cells may be traced, to a large extent at

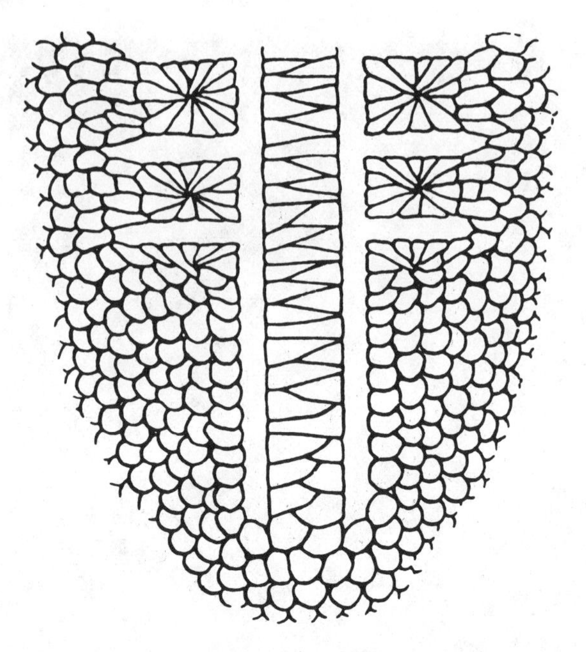

Figure 7.13. Diagrammatic illustration of the mesoderm segmentation. From Waddington (1956).

least, to the sclerotome and the neural crest. It is remarkable, however, that except for the most anterior part of the cranial neural crest, which produces nerve cells, the remaining part contributes to the formation of the skeleton of the cranium and the gill arches. These are typical mesodermal patterns of differentiation, and it is difficult to believe that we may here find the source of the mesenchyme cells, which cause the swelling of the head. Most likely the mesenchyme cells are derived from those free mesodermal cells which advance in front of the mesoderm (cf. fig. 7.11).

The various facts discussed here show that there is no hard and fast distinction between the neural crest (ectomesoderm or mesectoderm) and the true mesoderm. In passing it might be mentioned that in the anurans this swelling is not confined to the cephalic region, rather the whole trunk swells. This fact upsets morphogenesis to the extent that the anatomy of the skeleton deviates considerably from that distinguishing the normal vertebrate body plan (fig. 7.15).

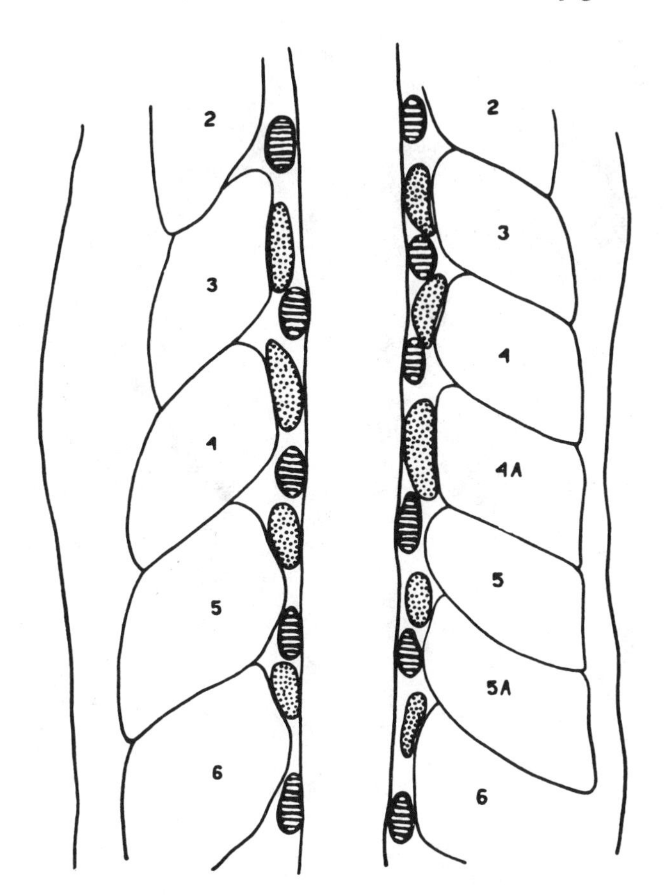

Figure 7.14. Experimental demonstration of the causal connection between the metamery of somites, the vertebrae, and the spinal ganglia. Stippled, vertebrae (neural arches); shaded, spinal ganglia. From Balinsky (1981), after Detwiler.

Only two more morphogenetic events shall be discussed, the formation of the median fin and the limbs. It appears that when most of the trunk neural crest cells have left their station, some cells remain locked in the dorsal median above the spinal cord. When they undergo differentiation, they become mesenchyme cells producing hyaluronate. The consequent swelling extends the epidermis, but instead of forming a semicylindrical extension, a gradually tapering dorsal fin is formed. Instrumental in this process are sf-cells suspended between the two walls of the fin.

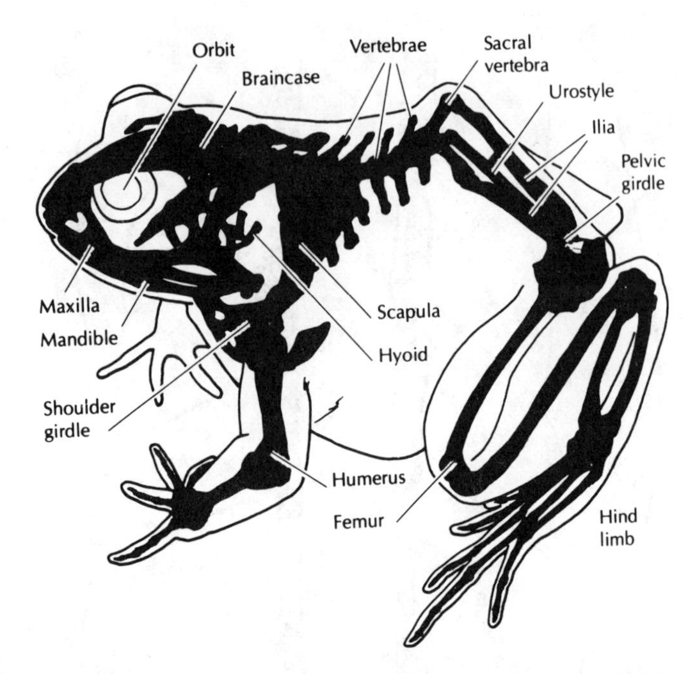

Figure 7.15. The anuran tadpole is morphologically unique by lacking a clear transition between head and trunk. The reason is, in part at least, that the swelling normally confined to the head occurs even in the trunk. As a consequence the integument is lifted off the somites, thus obliterating the morphogenetic basis for the formation of the ribs. The result is clearly shown in the skeletal anatomy of the adult frog. From Gans (1974).

In most tetrapods ribs are absent in the most anterior and the most posterior part of the trunk, thus creating the spatial requisite for the formation of front and hind limbs respectively. Various observations demonstrate that the absence of ribs is a precondition for the formation of the limbs; suffice it to mention that in general ribs are formed throughout the length of the trunk in the limbless tetrapods.

With the exception of the urodeles, limb formation begins with the appearance of ridges consisting of elongated cells in the epidermis, parallel to the length of the body. We do not know what causes this apparent cell differentiation, but since it occurs at the lower edge of the somites it might be presumed that these are somehow involved. The buds begin

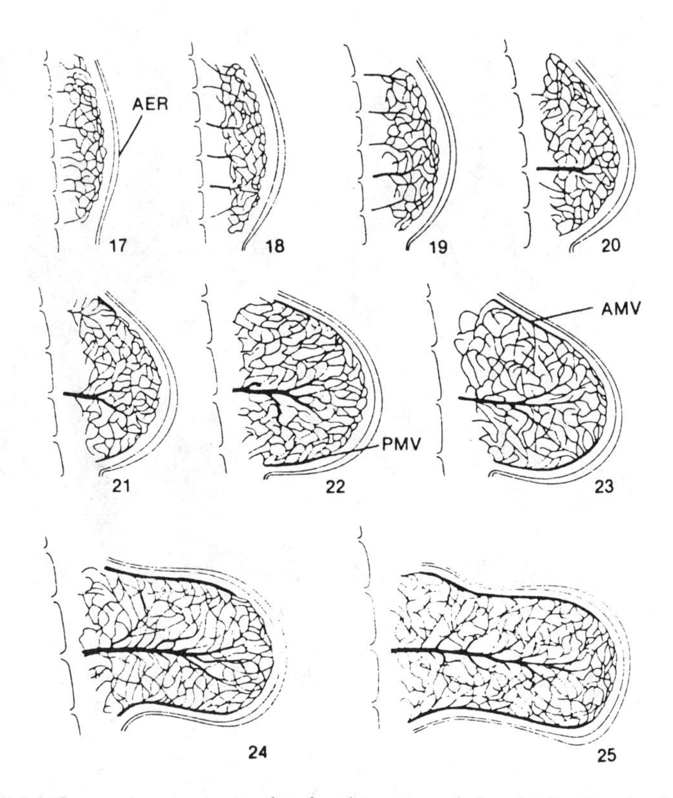

Figure 7.16. Successive stages in the development of the chick wing bud demonstrating the vascularization associated with the swelling of the bud. *AMV*, anterior marginal vein; *PMV*, posterior marginal vein. From Feinberg and Saunders (1982).

to swell, again the consequence of hyaluronate-producing mesenchyme cells (fig. 7.16). When the bud has reached a certain size, a condensation of cells is seen to take place. These cells begin to produce cartilage, thus forming a cartilagenous rudiment of the limb skeleton. The cells discussed so far are derived from the lateral mesoderm, whereas the cells forming the skeletal muscles bud off from the lower edges of the somites.

It might be envisaged that the typical deviations from the typical pentadactyl limb skeleton, found in some tetrapods (e.g. birds and horses), were the outcome of processes of differential growth. This is not the

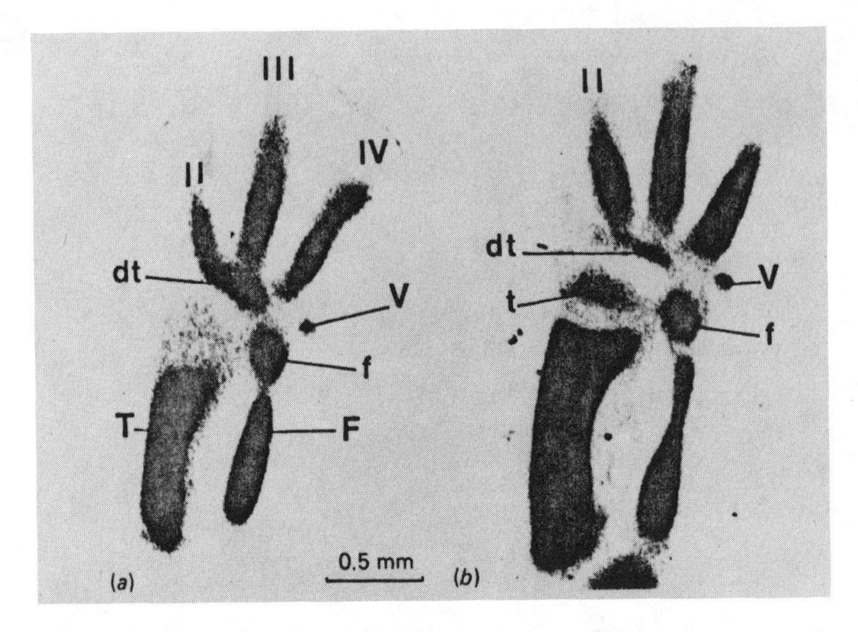

Figure 7.17. Autoradiographs of $^{35}SO_4$ uptake by the chick forelimb (stage 30) and hindlimb (stage 27/28). From Hinchliffe and Griffiths (1982).

case; rather, they are established at the outset, as may be seen by comparing the limb buds of the bird's wing and foot, and the horse's foot (figs. 7.17 and 18).

Differential growth

There are obvious differences between the embryos belonging to the various craniate classes at the end of the phase of form creation, but within the classes it is generally impossible to determine taxonomic affinity. Those enormous differences which obtain between, say, adult birds or mammals with respect to size and form have yet to be established. As far as the first parameter is concerned, it might be observed that it is correlated with the duration of incubation or pregnancy, the bigger the animal at hatching or birth, the longer time it takes to reach this point.

But it is also during this period that the various morphological peculiarities are developed, and this happens through differential growth, affecting partly the skeleton, as exemplified, by the giraffe, and partly the

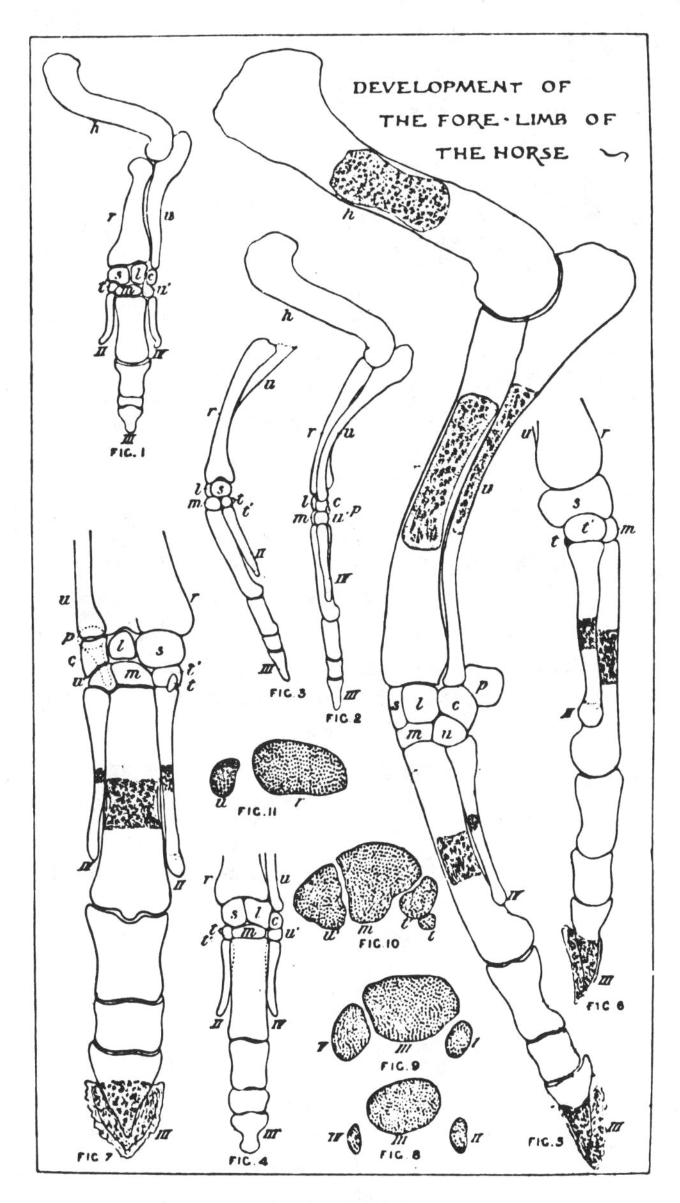

Figure 7.18. Various stages of the phase of differential growth of the forelimb of the horse. In the youngest stage, the skeletal anatomy is already established in detail. Thus, the subsequent changes can only be quantitative, not qualitative. From Ewart (1894).

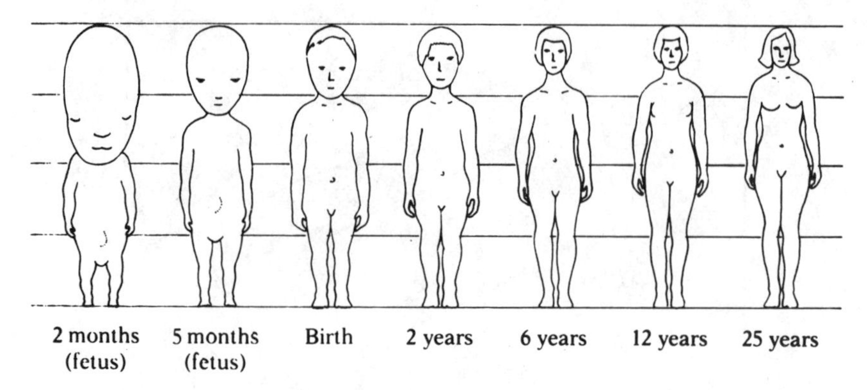

Figure 7.19. Allometric growth in *Homo sapiens*. From Keeton (1980).

soft parts of the body, such as the trunk of the elephant. It should be noted that cell death, and consequent suppression of morphological singularities, is of great importance during this phase.

The examples described here suffice to show that variation in the growth of the skeleton has been the most important agent in evolutionary change in most craniate taxa. In particular, in many cases the relative growth rates of the length and width of various bones have been the cause of evolutionary change. These two parameters are frequently, if not always, controlled by ossification processes. We know that the width of bones is influenced by different hormones, and it is possible that variations in these hormones may account for the observed morphological modifications, although other factors might also be involved.

Allometric growth

For obvious reasons the phase of allometric growth during ontogeny is perhaps the best understood. Allometric growth in our own species is amply recorded, and as shown in Fig 7.19 the most remarkable features are that in relative terms the head is very large at birth, and the limbs very short. The former feature seems to be common in birds and mammals, but not elsewhere in the craniates; it has reached the most extreme level in humans than in any other taxon. Conversely, in many instances the limbs of newborn mammals are relatively longer than those of adults.

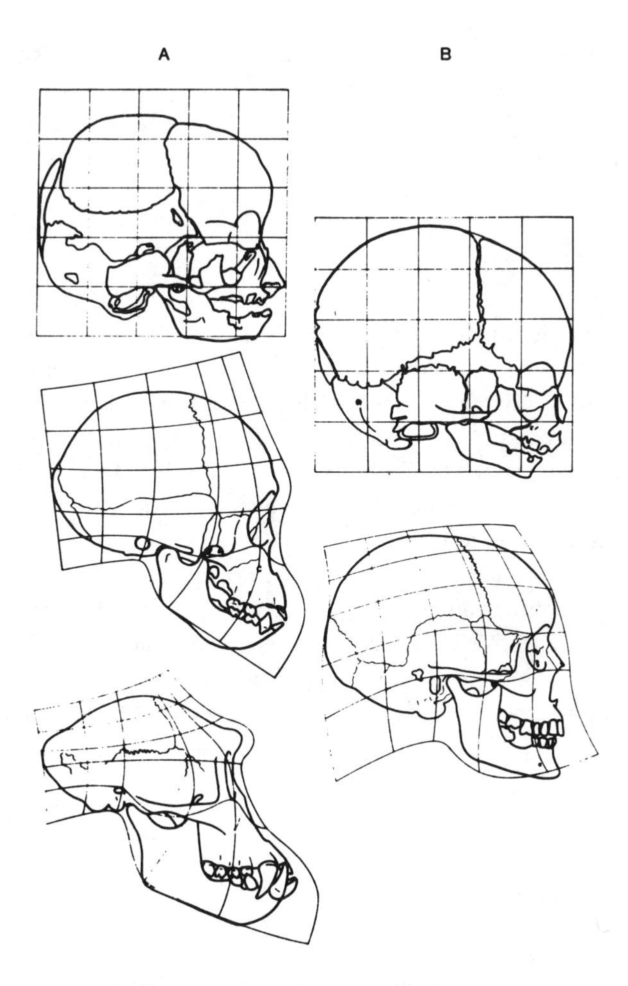

Figure 7.20. Postnatal allometric growth (A) in the chimpanzee and (B) in the human skull. In spite of the great initial similarity, there is a large difference between the adults, owing to a much more extensive allometric growth process in the ape. From Gould (1977), after Starck and Kummer.

A very well-known example of allometric growth is the great difference in head shape of adult chimpanzees and human beings (fig. 7.20).

It seems that birds, as compared to mammals, make greater use of the allometric growth phase. In birds distinguished by excessive adult traits such as very long beaks and legs, such features are established first during this phase. An exception to the rule is found in bats, in which the young are born with forelimbs having almost "normal" proportions. It is only during the early postnatal period that the forelimbs become transformed into wings (Yalden and Morris 1975).

CONCLUSION

I have defined "epigenetics" as "the study of the mechanisms responsible for the effectuation of ontogenetic development." If this definition is accepted, two further premises must be accepted for bringing "epigenetics" within range of the topic "Evolution and Systematics."

First, "systematics" must stand only for "phylogenetic classification." Second, it is necessary to accept the ideas of von Baer, according to whom the features distinguishing taxa to which a particular embryo belongs appear in a succession corresponding to subordination of the various taxa. If this assertion is accepted then ontogenetic development will constitute a recapitulation of the course of phylogenetic evolution. Furthermore, a comparative study of the epigenetic mechanisms responsible for ontogenesis in individual cases may thus serve as the empirical method to elucidate through which process evolution has been accomplished.

REFERENCES

Baer, K.E. von 1828. *Ueber die Entwickelungsgeschichte der Thiere*. Königsberg: Gebrüder Bornträger.

Balinsky, B.I. 1981. *An Introduction to Embryology* (5th ed.). Philadelphia: Saunders College Publishing.

Ballard, W.W. 1981. Morphogenetic movements and fate maps of vertebrates. *American Zoologist* 21:391–399.

Bellairs, R. 1979. The mechanism of somite segmentation in the chick embryo. *Journal of Embryology and Experimental Morphology* 5:227–243.

Britten, R.J. and E.H. Davidson, 1971. Repetitive and non-repetitive DNA sequences and a speculation on the origins of evolutionary novelty. *Quarterly Review of Biology* 46:111–138.

Brachet, A. 1921. *Traité d'Embryologie des Vertébrés*. Paris: Masson.

Castle, W.E. and P.W. Gregory, 1929. The embryological basis of size inheritance in the rabbit. *Journal of Morphology and Physiology* 48:81–102.

Darwin, C. 1859. *On the Origin of Species*. London: John Murray.

Ewart, J.C. 1894. The development of the skeleton of the limbs of the horse, with observations on polydactyly. *The Journal of Anatomy and Physiology* 28 (*NS* 8): 236–256.

Feinberg, R.N. and J.W. Saunders, 1982. Effects of excising the apical ectodermal ridge on the development of the marginal vasculature of the wing bud in the chick embryo. *Journal of Experimental Zoology* 219:345–354.

Gans, C. 1974. *Biomechanics: An Approach to Vertebrate Biology*. Ann Arbor: University of Michigan Press.

Gould, S.J. 1977. *Ontogeny and Phylogeny*. Cambridge: Harvard University Press.

Haeckel, E. 1866. *Generelle Morphologie der Organismen*. Berlin: Georg Reimer.

Haeckel, E. 1920. *Natürliche Schöpfungsgeschichte* (12th ed.). Berlin: Vereinigung wissenschaftlicher Verleger.

Hennig, W. 1966. *Phylogenetic Systematics*. Urbana: University of Illinois Press.

Hinchliffe, J.R. and P.J. Griffiths, 1982. The prechondrogenic patterns in tetrapod limb development and their phylogenetic significance. In B.C. Goodwin, N. Holder and C.C. Wylie, eds. *Development and Evolution*, pp.99–121. Cambridge: The University Press.

Holtfreter, J. 1943. Properties and functions of the surface coat in the amphibian embryo. *Journal of Experimental Zoology* 93: 251–323.

Hörstadius, S. 1950. *The Neural Crest*. London: Oxford University Press.

Jacobson, A.G. 1978. Some forces that shape the nervous system. In C.-O. Jacobson and T. Ebendal, eds. *Formshaping Movements in Neurogenesis*, pp. 13–21. Stockholm: Almqvist & Wiksell.

Janvier, P. 1981. The phylogeny of the Craniata, with particular reference to the significance of fossil "agnathans". *Journal of Vertebrate Paleontology* 1:121–159.

Keeton, W.T. 1980. *Biological Science* (3rd ed.) New York: Norton.

Lamarck, J.-B.-P.-A. 1809. *Philosophie Zoologique*. Paris: Dentu.

Løvtrup, S. 1974. *Epigenetics—A Treatise on Theoretical Biology*. London: Wiley.

Løvtrup, S. 1977. *The Phylogeny of Vertebrata*. London: Wiley.

Løvtrup, S. 1978. On von Baerian and Haeckelian recapitulation. *Systematic Zoology* 27:348–352.

Løvtrup, S. 1983. Epigenetic mechanisms in the early amphibian embryo: Cell differentiation and morphogenetic elements. *Biological Reviews* 58:91–130.

Løvtrup, S. 1984. The mechanism of the amphibian primary induction at the cellular level of organization. In A.-M. Duprat, A.C. Kato and M. Weber, eds. *The Role of Cell Interactions in Early Neurogenesis*, pp. 55–66. New York: Plenum Press.

Løvtrup, S. 1987. *La préformation et l'epigénèse in Hommage au Professeur Pierre-Paul Grassé*, pp. 87–98. Paris: Masson.

Løvtrup, S. and K. Hansson Mild. 1979: Limb bone robustness: Natural selection

or allometric growth? *Bulletin Biologique de la France et de la Belgique* 113:1–16.
Løvtrup, S., U. Landström, and H. Løvtrup-Rein. 1978. Polarities, cell differentiation and primary induction in the amphibian embryo. *Biological Reviews* 53:1–42.
Løvtrup, S., F. Rahemtulla, and N.-G. Höglund 1974. Fisher's axiom and the body size of animals. *Zoologica Scripta* 3:53–58.
Løvtrup, S., A. Rehnholm, and R. Perris. 1984. Induction of the synthesis of melanin and pteridine in cells isolated from the axolotl embryo. *Development, Growth and Differentiation* 26: 445–450.
Løvtrup-Rein, H., U. Landström, and S. Løvtrup 1978.Lactate—a suppressor of differentiation in embryonic cells. *Cell Differentiation* 7: 131–138.
Mattsson, M.-O., H. Løvtrup-Rein, and S. Løvtrup. 1987. Factors involved in the formation and stabilization of cell aggregates obtained from amphibian embryonic explants. *Cell Differation* (in press).
Mookerjee, S., E. M. Deuchar and C.H. Waddington, 1953. The morphogenesis of the notochord in Amphibia. *Journal of Embryology and Experimental Morphology* 1:399–409.
Nakatsuji, N. 1984. Cell locomotion and contact guidance in amphibian gastrulation. *American Zoologist* 24:615–627.
Needham, J. 1942. *Biochemistry and Morphogenesis.* Cambridge: The University Press.
Newport, J. and M. Kirschner. 1982. A major developmental transition in early Xenopus embryos: 1. Characterization and timing of cellular changes at the midblastula stage. *Cell* 30:675–686.
Painter, T.S. 1928. Cell size and body size in rabbits. *Journal of Experimental Zoology* 50:441–453.
Raff, R.A. and T.C. Kaufman. 1983. *Embryos, Genes and Evolution. The Developmental-Genetic Basis of Evolutionary Change.* London: Macmillan.
Roff, D. 1977. Does body size evolve by quantum steps? *Evolutionary Theory* 3:149–154.
Rugh, R. 1948. *Experimental Embryology: A Manual of Techniques and Procedures.* Minneapolis: Burgess.
Sewertzoff, A.N. 1931. *Morphologische Gesetzmässigkeiten der Evolution.* Jena: Gustav Fischer.
Stern, R. N. and A. B. Kostellow. 1958. Enzyme induction in dissociated embryonic cells. In W.D. McElroy and B. Glass, eds. *A Symposium on the Chemical Basis of Development,* pp. 448–453. Baltimore: Johns Hopkins Press.
Vogt, W. 1929. Gestaltungsanalyse am Amphibienkeim mit örtlicher Vitalfärbung. II Teil. Gastrulation and Mesodermbildung bei Urodelen und Anuren. *Wilhelm Roux' Archiv für Entwicklungsmechanik* 120:384–706.

Waddington, C. H. 1956. *Principles of Embryology*. London: George Allen and Unwin.

Weston, J. A. 1970. The migration and differentiation of neural crest cells. *Advances in Morphogenesis* 8:41–114.

Willmer, E. N. 1960. *Cytology and Evolution*. New York: Academic Press.

Yalden, D. W. and P. A. Morris. 1975. *The Lives of Bats*. Newton Abbott, England: David and Charles.

Index